Freshwater Ecology

Alison Leadley Brown, M.A., F.I. Biol.

Heinemann Educational Books

Heinemann Educational Books Ltd
22 Bedford Square, London WC1B 3HH

LONDON EDINBURGH MELBOURNE AUCKLAND
SINGAPORE KUALA LUMPUR NEW DELHI
IBADAN NAIROBI JOHANNESBURG
PORTSMOUTH (NH) KINGSTON

ISBN 0 435 60622 0

First published 1987

Typeset in 10 on 11½ Plantin by
Image Communications Ltd.
Printed and bound in Great Britain by
Richard Clay Ltd, Bungay, Suffolk.

Preface

The intention in writing this book is to give an introduction to the study of the ecology of fresh waters. Several decades ago such a study would have relied upon a straightforward account of the structure and function of freshwater animals and plants without reference to how they enabled an organism to exist in its chosen habitat. Nowadays the word 'ecology' has become common usage in many contexts but often with different connotations. In the biological sense ecology could be defined as the experimental analysis of distribution and abundance or, more simply, the study of living organisms in relation to their environment.

Any ecological study will include many branches of science. The central purpose for the freshwater biologist will be to interpret principles and to show how these principles can be applied to the study of different kinds of fresh water.

Ponds, lakes, streams and rivers are an integral feature of our landscape. They play important roles in a number of human requirements such as water supply, waste disposal, and leisure, all of which make them vulnerable areas and in need of protection by effective management. Indeed, management has provided a strong impetus to the study of fresh water in all its forms and at all levels.

The factors affecting the spatial distribution of freshwater animals and plants is central to any investigation. Here, the author has attempted to illustrate theoretical considerations by examples of fieldwork which it is possible to carry out in still and flowing water.

While an understanding of the methods of statistical analysis of quantitative results is necessary, especially with the increasing use of computers, nothing can replace observation at first hand of the behaviour of organisms in the context of their environment and the careful recording of such observations.

If this book serves to encourage students to embark on a fresh water project and to make their own studies of the structure and behaviour of animals and plants living in a chosen habitat or microhabitat, then one of its main objects has been achieved.

My grateful thanks are due to Professor W. H. Dowdeswell for his advice throughout the writing of the book and for reading the final manuscript. I should also like to thank Professor David Nichols for Plate 3.1 and Andrew Cooper for Plate 11.3. The line drawings throughout and all other photographs are by the author.

Contents

Introduction

It has been generally accepted that life began in the sea and that colonization of the land took place by invasion from the sea. What more natural, then, than to suppose that this took place via fresh water. However, this may not in general be so for the invaders of fresh water, be they from the land or the sea, are immediately presented with a number of problems and no single species has solved them all. While the sea offers the most constant medium in which to live, fresh water is liable to greater variation of temperature and salt concentration, especially after heavy rain.

The degree of adaptation imposed on any invaders of fresh water varies from practically no alteration in structure, typified by some species of water beetle, to a considerable modification of anatomy and physiology required by species inhabiting the swift currents of an upland stream and by those living in mud where there may be a complete absence of oxygen.

Many of the physical factors of fresh water have proved an effective barrier to all but a few organisms. Successful species include a few fish and molluscs such as Jenkins' spire shell and the freshwater shrimp, *Mysis relicta*.

An examination of the animal phyla to which freshwater species belong shows that invasion of fresh water has largely been from the land and this is the case for all freshwater plants. Some animals, normally resident in the sea, migrate for breeding purposes to fresh water: the salmon and marine lamprey, for instance. The reverse is true for the common eel which migrates to the sea in order to breed.

In any medium, life for both plants and animals is a constant struggle. Besides having to adapt to the physical environment, predators, parasites, and other competitors have to be contended with. Yet, compared with a terrestrial existence, fresh water offers a relatively stable environment.

Plants and animals live in breeding populations and not in isolation. Such populations, or more often several populations, live as a community. The words 'population' and 'community' are often used loosely in describing an ecosystem and the confusion can become greater when the composition of a microhabitat is investigated. Uniformity of conditions within a microhabitat, the warm outflow from the cooling towers of an electricity generating station for instance, can mean a restriction in the number of species which are adapted to live there. In these cases it is open to question whether such a group of organisms can be described as a community.

It must not be forgotten that the behaviour of an organism is an integral part of the way in which that organism is adapted to live in a particular habitat. Whereas the spatial distribution of plants is governed to a large extent by their physical environment, that of animals is a much more complex selective process. This inevitably involves us in a study of the animal's behaviour in order to interpret the reasons for its distribution.

The reader will find that the first few chapters are devoted to a description of the important factors, chemical and physical, existing in still and flowing water. Later chapters set out to interpret the ways in which plants and animals have adapted to these conditions. Some organisms are capable of living and breeding in both running and still waters and so a clear-cut distinction between the two cannot be made. Nevertheless, the study of a community in either or both localities can only result in rewarding experiences.

The selection of a particular project is not always easy and preparation beforehand is essential if time is not to be wasted in recording unnecessary and irrelevant information. The availability of a site for study, the time of year, and, above all, the defining of the boundaries of a study are all important. After these criteria have been decided, there remains the keeping of accurate records in such a way that they can be interpreted. Finally, the methods employed for the interpretation of results must be decided upon.

In a book of this length it is clearly impossible to include all the background information and statistical methods for the interpretation of results. Use should therefore be made of the Bibliography at the end of the book which is divided into sections. The section headed 'Ecological methods' includes several books which will be of help in the practical investigation of a habitat and in methods of recording and interpreting results.

1 The economy of fresh water

This book is concerned with fresh water and all the animals and plants that live in it. The study of fresh water and its ecology can provide endless opportunities for field work and some of the most enjoyable experiences that biology can offer. But first we must be clear as to what we mean by the word 'ecology'.

Living organisms seldom occur alone but are usually found in numbers together. Sometimes these aggregations are all of the same species and they are therefore able to interbreed quite freely; such a group is known as a **population**. More often, however, the occupants of a particular locality or habitat comprise a number of different species which depend upon one another for food and shelter and so forth to varying degrees. Several interacting populations are called a **community**.

Ecology is the term used to describe the relationships between living organisms and their environment and an **ecosystem** is the relationship between a community or a number of communities and their environment.

Fresh water offers a variety of ecosystems, depending upon the type of water and the organisms present. The still waters of lakes and ponds offer different habitats, depending upon a number of factors, among which are depth and the chemical and physical quality of the water. The running water of rivers and streams offers yet another kind of environment which is dependent upon characteristics such as the speed of the current and the nature of the stream bed.

Communities of living organisms existing within a locality are dependent upon one another for food. Some depend upon others and so there must be a balance between those which produce and those which consume. Survival depends upon this balance, which poses the question: who are the producers?

Producers and consumers

Plants are capable, by the process of photosynthesis, of fixing energy from the sun and this energy is incorporated into various compounds containing carbon. This is used, in part, by the plants themselves in respiration and by those organisms which consume them. Green plants are, therefore, the **producers**.

Because all the energy entering an ecosystem is due to the activities of green plants they are called **primary producers** and **primary production** is the material they produce. The amount of production (**productivity**) is the quantity of material produced per unit time. This will depend upon the amount of light energy and the quantity of nutrients available, both of which are essential for photosynthesis. In taking up nutrients and incorporating them into their tissues, plants deplete the pool of essential substances in the environment which must be restored in other ways.

Organisms which consume plants are called **herbivores**, those which consume animals are **carnivores** and those which eat both plants and animals are **omnivores**.

Herbivores eating plant food are the **primary consumers** while those capturing and feeding on the primary consumers are the **secondary consumers**. Herbivores, carnivores, and omnivores, during the course of their growth and multiplication, create the **secondary production**. But because energy losses in animal respiration are greater than those incurred in plant respiration, secondary production will always be less than primary production and that of carnivores will be less than that of herbivores.

Decomposers

Not all the energy fixed by plants is available to the herbivores because many plants die before they are eaten and some plant tissues may be indigestible. In autumn, fallen leaves add yet more plant material which is not consumed. So the waste organic matter will comprise the uneaten and indigestible plant material and the faeces of the animals present, together with their dead remains. Bacteria and fungi are the decomposers responsible for breaking down this waste matter to release essential elements such as nitrogen and phosphorus which are returned to the ecosystem and recirculated. The decomposers themselves and their food source may be consumed by **detritivores** (detritus feeders) which are often eaten by carnivores.

Food relationships.

It is clear from the above description that in any ecosystem there is a close relationship between eater and eaten. Such a combination is a feeding or **trophic** relationship, starting with plants which form the first trophic level. Plants are eaten by animals which form the second trophic level.

We could give as an example of the simplest trophic relationship grass eaten by cows and this chain could then be enlarged one step further by adding that cows are killed and eaten by human beings who would then form the third trophic level.

In a simple case such as the following:

green algae → water flea → roach → pike

the pike becomes the fourth trophic level. This feeding sequence is an example of a simple **food chain** but there may be many different species of herbivore feeding on the green algae and water fleas are the food of many species of carnivore. Several larger carnivores may feed on the roach. The simple food chain then immediately becomes quite a complicated **food web** comprising a number of **grazing food chains** (living plants → herbivores → carnivores) as well as **detritus food chains** (detritus → detritivores → carnivores).

The amount of food available at any one time to the primary consumers will depend upon the amount of living plant material present, or its **biomass**. (The **standing crop** is the biomass present at a particular moment.) That available over a period of time will depend on the rate at which it is produced i.e. **productivity.**

Diversity of species within an ecosystem

Some of the net production in nearly all ecosystems is consumed by animals, whether they are large, such as cattle, or microscopic, feeding directly upon living plants. These are the primary consumers or grazing herbivores and, as we have already mentioned, they form the food of the carnivores. The flow of food through this channel is the **grazing food chain**. The rest of the net production is destined to be consumed by the detritivores as dead organic matter such as dead leaves, twigs, algae, and so on. This is the **detritus food chain**.

Different ecosystems, and even parts of the same system, vary according to the amount of energy passing along the grazing as opposed to the detritus food chains. In heavily grazed pastureland, for instance, 50 per cent or more of the annual net production may pass down the grazing food chain, although not all the food eaten will be actually assimilated. Some, such as the faeces produced, will be diverted to the decomposer route. The contribution made to the ecosystem by the grazers will depend on the amount of plant material removed from the standing crop as well as on the amount of energy contained in the food consumed.

In contrast to the ecosystem of grazed pasture, less than 10 per cent of the net production of a fertile intertidal salt marsh is consumed by grazing herbivores which, in this case, are mostly insects. At least 90 per cent passes through the detritus food chain, shellfish, snails, and small crustaceans being the detrital feeders and the dominant salt marsh invertebrates.

In considering some general principles we find that ecosystems vary as to the number of different species they support and it is often assumed that an ecosystem with a large biomass (B) will also have a high productivity (P). But this is not always true since several factors must be taken into account, such as the size of the organisms present and their longevity. Large plants or large animals, for example trees and large herbivores such as elephants, will accumulate a large biomass during their long lives but will have a low $P:B$ ratio when this is calculated on an annual basis, because although such large organisms can build up a large biomass, much of the primary production is taken up in order to maintain their respiration.

Long-lived slow-growing forms will build up a biomass to the maximum that their environment can support. On the other hand an environment which does not favour such species will offer opportunities for short-lived species which, when conditions are favourable, grow and reproduce fast. For instance large numbers of water fleas can occur in a pond at certain times of the year when nutrients are abundant and temperature and light intensity are greatest, favouring the rapid multiplication of algae upon which the water fleas feed. Less of the primary production will be needed to maintain a high biomass over long periods, so the $P:B$ ratio tends to be high. Under these conditions there will be a fluctuating biomass.

The ways in which these general principles apply to freshwater communities are discussed in subsequent chapters.

2 Fresh water as an ecological environment

All freshwater habitats are dynamic systems involving close interaction of the organisms with the physical and chemical conditions prevailing. The organisms themselves are an important part of the environment, often capable of exerting a profound influence upon the habitat.

Biotic factors are those which result from the interaction of living organisms with one another and these, together with chemical and physical factors, go to make up a freshwater ecosystem. All influence the distribution of the organisms living there, interacting to cause fluctuations in the composition and size of populations.

Because the various factors which are described below seldom act in isolation, it is often difficult to separate the effect of one from another. It is, therefore, not usually possible to attribute the adaptations of structure and behaviour made by an organism to one particular factor when, in reality, the reactions may be due to several acting together. For instance, the amount of dissolved oxygen in a body of water is dependent both upon the temperature of the water and the atmospheric pressure, two factors which may each produce different reactions.

In a brief survey of this kind there is not space to make an exhaustive study of the whole complex of factors affecting freshwater organisms. Specific examples will be given and reference should also be made to the Bibliography.

Physical characteristics of fresh water

Light and radiant energy

In most freshwater ecosystems light is the major source of energy. The amount of light reaching the surface of a body of water will depend on the amount of cloud cover, the altitude, geographic position, and the season of the year.

As the rays of light meet the surface of the water some are reflected, others are absorbed. The amount of reflected light will depend on the angle of incidence (Fig. 2.1). The energy reflected from the surface varies, therefore, not only diurnally but also seasonally.

Radiant energy diminishes with depth until there is virtually none below a certain level (Fig. 2.2), this depth being dependent upon the amount of dissolved and suspended solids

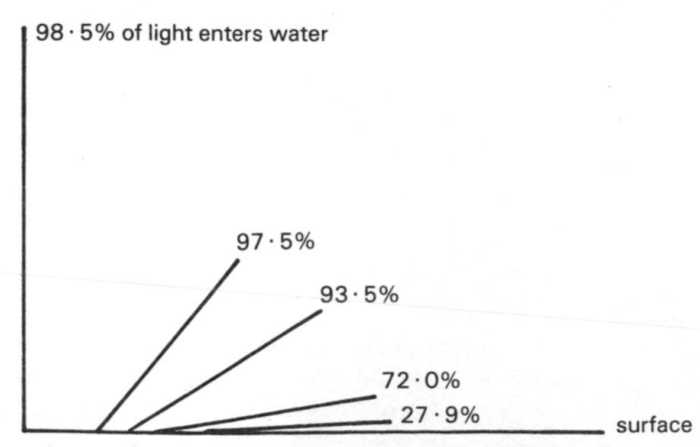

Fig. 2.1 The percentage reflection of light from the surface of water depends on the angle of incidence

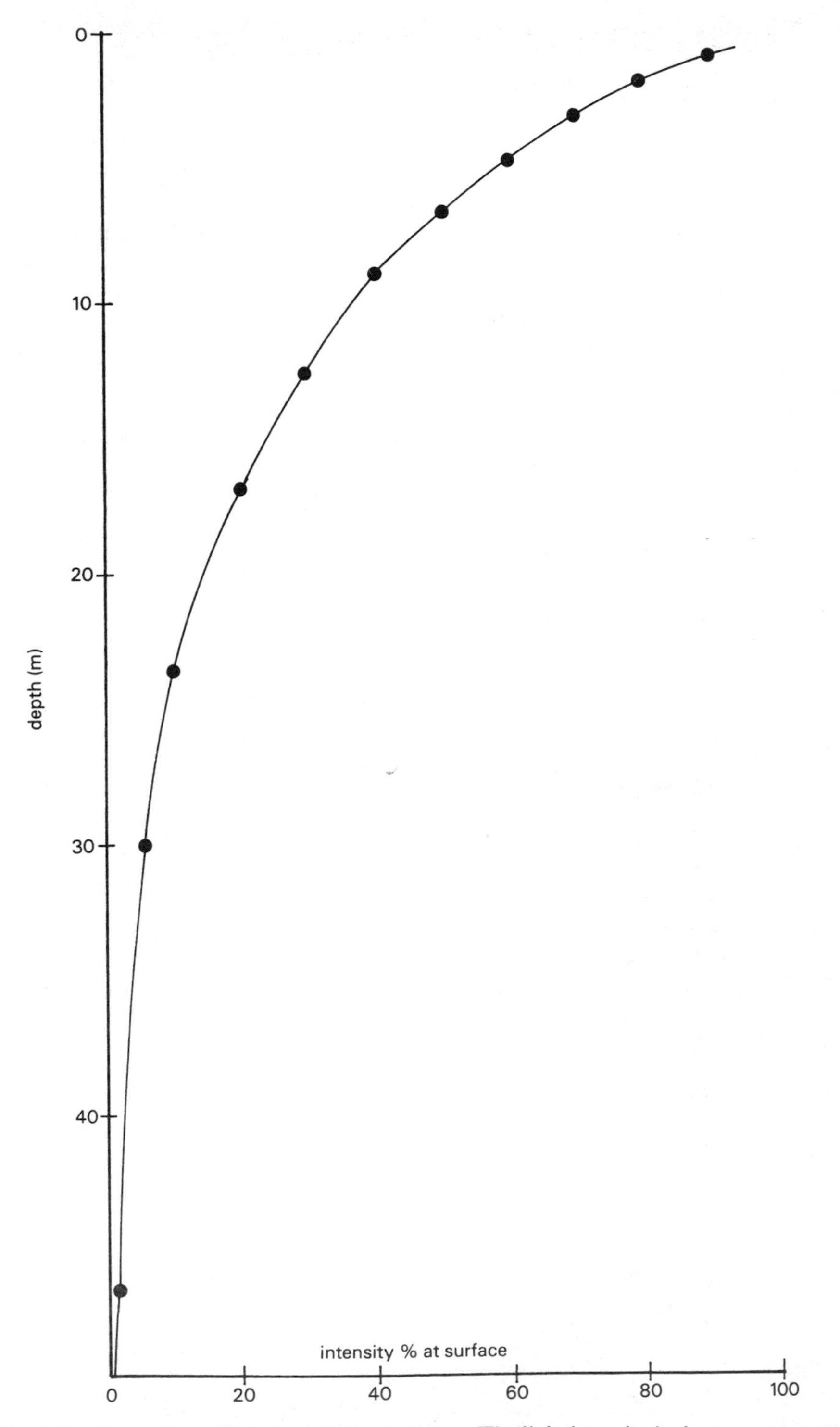

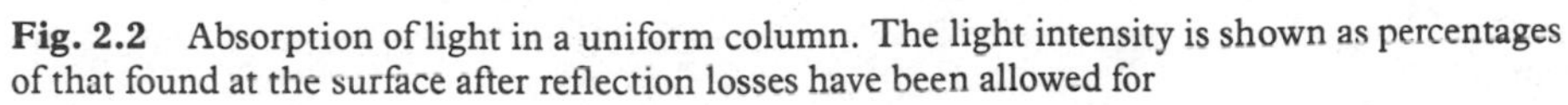

Fig. 2.2 Absorption of light in a uniform column. The light intensity is shown as percentages of that found at the surface after reflection losses have been allowed for

present in the water. The longest and shortest wavelengths of the light spectrum — the reds and blues — are absorbed most rapidly, while the middle wave bands — the yellows and greens — penetrate the deepest. Red and orange rays are those most effective for the photosynthetic activities of green plants. Hence the distribution in depth of underwater plant communities will be governed to a large extent by the degree of radiant light penetration. A Secchi Disc is a simple method of comparing the transparency of different waters (p. 11, Fig. 2.6 and [2.7].

Ice is another factor controlling the amount of light penetration. Thin, so-called 'black' ice allows a high degree of penetration but where ice is thick and rough or covered by snow, the amount of light reaching the water below is greatly reduced.

Density and temperature

At 0°C and atmospheric pressure (760 mm Hg) the density of pure water is 700 times that of air. This means that the bodies of aquatic organisms need much less structure to support them than their terrestrial relatives. There is a reduction of skeletal structure in aquatic plants, especially in their underwater stems and leaves (see Chapter 5).

Most liquids increase in density as the temperature decreases, but with water the temperature/density relationship is reversed. Water reaches a maximum density at 3.94°C, below which the density decreases gradually to its freezing point (0°C). Below 0°C the decrease in density increases sharply (Fig. 2.3).

This change in density with temperature is of great importance to freshwater organisms and has a considerable effect on their distribution patterns. For instance, a 1°C change in temperature at 24°C means a density change many times more than at 4°C. Planktonic organisms will therefore tend to sink more rapidly at higher temperatures than at lower ones. Also, since aquatic animals, when moving about, are faced with a much greater resistance than their terrestrial counterparts, they have to expend more energy in doing so.

All aquatic invertebrates and fish are poikilothermic; that is, their body temperature varies with that of the surrounding water. In general the effects of lowering the temperature will be to slow down their life processes (digestion, respiration, movement, etc.) Many freshwater organisms are able to adjust to changes in density and temperature and can detect changes of as little as 0.2°C. Body colour can play a part in some invertebrates. Thus the adults of all winter-emerging stoneflies are black, enabling them to absorb the maximum amount of radiant heat, and dark colours which assist in camouflaging them against predators such as fish are found among the bottom-living species such as dragonfly nymphs.

Water movement

Wind is the prime factor to be considered in the movement of the surface waters, particularly in large lakes. Wind can either produce surface currents or, by driving the surface water towards the end of a lake, induce currents beneath the surface flowing in the opposite direction. The effect of surface waves decreases rapidly with depth and rarely penetrates deeper than 20 m. Nevertheless, this can have a profound effect on the distribution of organisms in the open water.

A different type of water movement is produced by convection currents which act in a vertical direction and are usually the result of sudden cooling of the water. In running water, currents have a different significance. Both these forms of movement and their effect upon the spatial distribution of organisms will be described in Chapters 4 and 10.

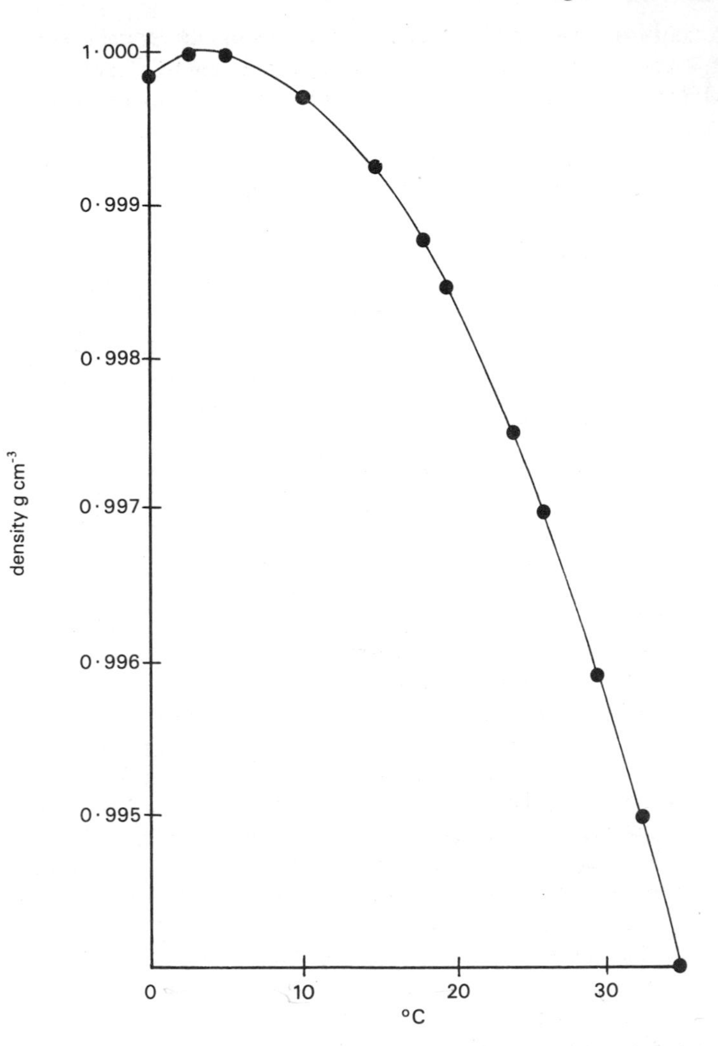

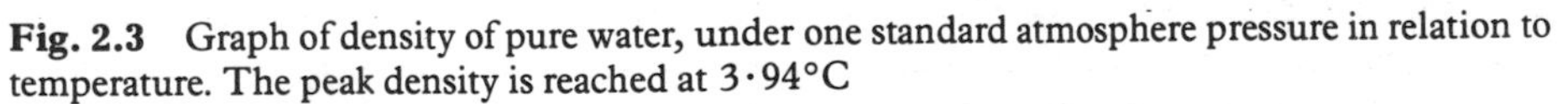

Fig. 2.3 Graph of density of pure water, under one standard atmosphere pressure in relation to temperature. The peak density is reached at 3·94°C

Surface tension

Another physical factor of water that is important to the community of animals and plants which live on the surface is surface tension, caused by the attraction of water molecules at the surface.

While the surface film is used by light insects such as pond skaters and water crickets to skate upon and by other creatures such as snails and flatworms to crawl underneath, it can act as a barrier to many air-breathing animals. These must adopt various means of penetrating the film in order to obtain a supply of air (Chapters 6 and 11).

Chemical characteristics of fresh water

All bodies of fresh water, whether lakes, ponds, rivers, or streams, contain chemicals of various kinds, either in solution or suspension. The concentration varies according to the nature of the substrate and that of the catchment area from which the water derives. Water occurring as surface run-off from fields which have been sprayed with chemicals containing nitrates and other substances can drastically alter the composition of any body of water and cause eutrophic conditions by the presence of excess minerals.

It must not be forgotten that rain water is never 'pure' for it contains appreciable amounts of calcium, potassium, sodium, magnesium, chloride, and sulphate as well as dissolved oxygen, carbon dioxide, and nitrogen from the atmosphere. Sea spray, which can be carried into the air for considerable distances, contributes sulphate in particular as well as sodium and chloride ions, while calcium, magnesium, potassium, nitrate, and phosphate are contributed by dust distributed by air currents.

In open sandy soils most water enters bodies of fresh water as sub-surface water (p. 00 and Fig. 7.1). The concentration of the dissolved solids in natural waters varies greatly, an average value being about 100 mgl^{-1}. More than this amount indicates drainage from sedimentary rocks, while in water containing less than 50 mgl^{-1} of solids, the parent rock is probably igneous. Communities of organisms present in a body of water contribute, by their own life processes, death and decay, in no small measure to the variation in the chemistry of the water in which they live.

Oxygen

To all but those organisms capable of living in conditions of low oxygen concentration, oxygen is essential to their life processes.

Atmospheric air, at normal temperature and pressure, contains 210 ml l^{-1} of oxygen, 780 ml l^{-1} of nitrogen, and small amounts of other gases including carbon dioxide. Oxygen is absorbed at the surface of the water but the degree of absorption is dependent upon the temperature and air pressure. The amount of oxygen that a given volume of water will hold in equilibrium falls as the temperature rises. The concentration of oxygen is expressed as the percentage of what it would be if the water were saturated with air at normal pressure (Fig. 2.4). Methods of measuring oxygen content of water and a table of the solubility of oxygen at different temperatures are given in Appendices 2A and 2B (pp. 14 and 15).

Figure 2.4 shows that even when water is saturated with oxygen it contains very little of the gas. At 5°C and normal air pressure, one litre of water contains only 8.9 ml oxygen, and at 20°C it contains 6.4 ml. With increasing altitude, air pressure becomes lower and the solubility of oxygen is also reduced. But an increase in altitude means a lowering of

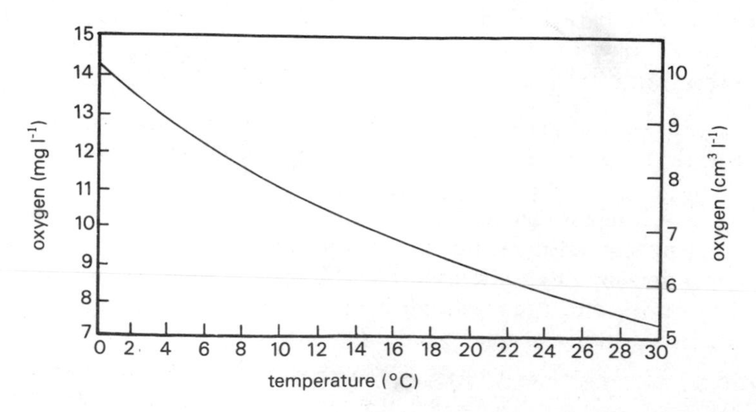

Fig. 2.4 The oxygen content of water saturated with air at normal pressure (740mg/Hg) (from Mortimer, 1956)

temperature. This increases the solubility of oxygen, thereby to a certain extent offsetting the effect of altitude on oxygen solubility. However, in the deeper waters of lakes at high altitudes there will probably be low concentrations of oxygen in solution at equilibrium.

The influence of temperature is closely associated with the consumption of oxygen by poikilothermic organisms. Their rate of metabolism, and therefore of oxygen consumption, increases with temperature by 10 per cent or more with every 1°C rise in temperature, so that as the temperature rises an oxygen shortage can easily be created.

The amount of oxygen present also influences many processes taking place, not least of which is organic sedimentation. Bacterial breakdown of plant and animal remains will be accelerated where adequate oxygen is available, while in conditions of oxygen deficiency, decay will take place much more slowly with a consequent lowering in the rate at which mineral products are returned to the environment.

Carbon dioxide and hydrogen ion concentration

Decomposition of organic matter, as well as the respiration of the living organisms present in fresh water, produces carbon dioxide which readily combines with water to produce carbonic acid (H_2CO_3). The carbonic acid dissociates to produce hydrogen (H^+) and bicarbonate (HCO_3^-) ions.

In pure water at 25°C, the concentration of H^+ ions is 10^{-7} moles l^{-1}. We use the log of the number of H^+ ions to describe the hydrogen ion concentration of the water, which is known as its pH. Water is acid if the pH value is less than 7, neutral if the pH value is 7, and alkaline when it is more than 7.

During daylight when the plants are actively photosynthesising, the pH may be greatly increased by the removal of carbon dioxide. Conversely, the production of large amounts of carbon dioxide and other organic acids resulting from decomposition lowers the pH, the water becoming more acid. During the hours of darkness when photosynthesis ceases, the carbon dioxide concentration rises due to respiration. Figure 2.5 shows the relationship between hydrogen ion concentration and the amount of free carbon dioxide, or that which is not combined as carbonate. Calcium carbonate, the result of the combination of CO_2 with calcium salts, causes the flocculation of colloidal humus and therefore an increase in the size of particles deposited, which improves permeability and aeration of the solids accumulating on the bottom. This favours the more rapid decomposition of organic matter.

Peaty moorland pools can have a pH of 4.0 or even less, while water derived from chalky soils will be alkaline, with a pH of 8 or more. The acidity or alkalinity of flowing water tends to remain fairly constant, the pH depending upon the source of water. That derived from run-off of bogs or dense forest litter will have a low pH, while water coming from a chalk spring will tend to be alkaline.

Dissolved minerals

The nature and amount of soluble mineral salts present in any body of fresh water will depend upon the geology of the surrounding land, since they enter in solution as drainage water. Mineral nutrients are also regenerated and released in solution as the result of the decaying processes brought about by bacterial and fungal action.

The main substances present in solution are chlorides, carbonates, phosphates, sulphates, and nitrates usually in combination with sodium, potassium, magnesium and calcium. Acid (soft) waters usually have low concentrations of calcium, magnesium, carbonate and sulphate. Conversely, hard waters contain high concentrations of these substances.

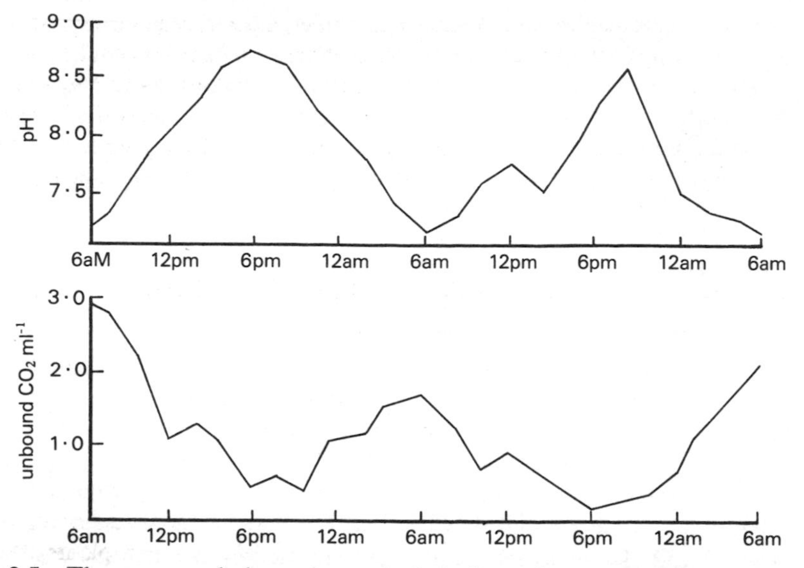

Fig. 2.5 The upper graph shows changes in the hydrogen ion concentration of surface water in a freshwater pond over a two-day period during late summer. The lower graph indicates the changes in carbon dioxide concentration which increases during the dark portion of the forty-eight-hour interval owing to an interruption of photosynthetic activity (from Knight, 1965)

Nitrogen, mostly in the form of ammonia and nitric acid, comes in small amounts from rain. As the water percolates through the soil, increasing amounts of nitrogen in the form of ammonia will be added as the result of the breakdown of animal and plant proteins, which are rapidly changed to nitrate by bacterial action. Some species of blue-green algae as well as bacteria are able to fix atmospheric nitrogen, which further enhances the supply.

Although often small, the supply of nitrates is important as a component of all living cells. In eutrophic waters almost all the nitrate present will be utilized by the green plants.

Phosphate, although important to all organisms, usually occurs in small quantities and is derived chiefly from sedimentary rocks in the catchment area. Much of the insoluble phosphate (ferric phosphate) is precipitated into the bottom sediments to remain there until anaerobic conditions occur, when the ferric iron is reduced to the soluble ferrous state, thus liberating phosphate. This is then available once more to the plant populations. In natural waters there is normally very little dissolved iron although more may be present in acid waters.

Calcium salts are probably the most important constituents of fresh water because the nature and size of populations of animals and plants chiefly depend upon the amount of calcium present. Calcium is also an essential constituent of skeletal structure and plays an important part in governing the permeability of cell walls as well as determining the diversity of the organisms present. Hence the wide range of animals and plants to be found in chalk streams and the comparatively restricted number of species in acid, upland streams.

Organic matter

Until quite recently, organic matter, present in most fresh water ecosystems, was given scant recognition. Now it is known to be of considerable significance as a source of carbon and other nutrients.

By the growth, reproduction and eventual death of organisms, organic material is made available by the various processes of decay, much of it being returned to the ecosystem to be recycled. Material derived in this way is often termed **autochthonous** and organic matter which originates outside the system, such as fallen leaves and other vegetation, **allochthonous.** A river, canal or lake heavily overshadowed by deciduous trees will receive a lot of dead foliage which is degraded by detritivores of various kinds, resulting in the release of soluble organic matter.

Some methods of measuring the characteristics of fresh water

Any ecological investigation will involve making measurements and recording these measurements. Before embarking upon the study of a particular freshwater habitat, consideration should be given to which environmental factors are involved. Measurements should be recorded in order to show how a factor or factors affect a community of organisms or individual species living in that habitat. Selection of an ecological study will, therefore, involve careful planning beforehand so that time is not wasted in unnecessary recording.

The four most important factors affecting freshwater animals and plants are light, temperature, oxygen, and pH, all of which have been described in this chapter. It must be remembered that in many cases they do not act in isolation, one influencing the other. Below are described some of the ways in which these factors can be measured.

Light

Light is most important because of its diverse effects, both direct and indirect, in influencing the distribution and numbers of organisms, particularly plankton, in a body of fresh water. Difficulties arise in measuring light, for light intensity is constantly varying. Floating material causes turbidity, reducing the penetration of light rays and thereby the rate of photosynthesis of plants growing beneath. A **Secchi disc** will give a determination of the depth of visibility, useful for purposes of comparison. Figure 2.6 shows the construction of the disc suspended by four cords joined to a calibrated line. The disc itself should measure approximately 200 mm in diameter and can be painted white, or in alternating black and white quarters with a weight suspended centrally so that it floats horizontally. To determine visibility the disc is lowered slowly into the water until it just disappears from sight, and the depth recorded. It should then be lowered for a further 0.5 m and raised slowly until it reappears, this depth being also recorded. The average of the two readings is the visibility.

By operating the disc in several regions of a pond a comparison of visibility can be made related to the distribution of plants in these areas.

For more accurate measurement of light intensity a **photovoltaic cell** must be used [2.2, 2.7].

Temperature

Changes of temperature in a pond or lake seldom exert their influence on the distribution or the rate of growth of organisms on their own, but in concert with other factors such as light. Temperature varies not only with depth but diurnally and seasonally, and changes must be measured. There are various types of thermometer which can be used. Ordinary glass mercury thermometers are unsuitable because of their fragility and the fact that readings are

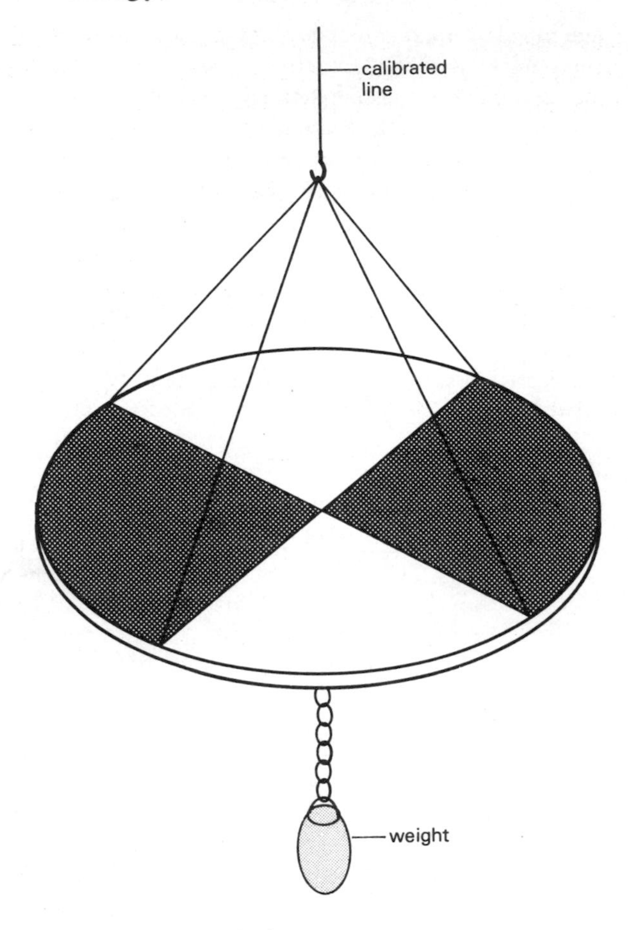

Fig. 2.6 Secchi disc

difficult to take at any depth below the surface. A **thermistor** is probably the most suitable and can be purchased or constructed in the laboratory [2.3].

Surface tension

The meniscus at the surface of water acts as a physical barrier between water and air to many invertebrates (Chapters 6 and 11). The surface dwellers are mostly light-bodied insects which also possess a covering of water-repellent (hydrofuge) hairs. Tension of the particles of water forming the meniscus can be lowered by adding drops of detergent.

Oxygen

Not only do living organisms adjust to physical and chemical changes in their environment but are themselves capable of bringing about changes, particularly in the amount of dissolved gases, carbon dioxide and oxygen.

In a well-illuminated pond in high summer the amount of oxygen at the surface may exceed saturation and bubbles of the gas may actually be present.

The amount of dissolved oxygen present in a pond at different depths and at different times of the year is an important factor in the distribution of invertebrates and is closely related to their methods of respiration. Those which live in the anaerobic conditions of mud, for instance the larvae of certain species of midge (*Chironomus*), possess haemoglobin which increases their ability to take up oxygen.

Various methods can be used to measure the amount of dissolved oxygen in water, the standard Winkler method (Appendix 2B) still being the one in common use. A modification of the Winkler method, using simple plastic syringes, has been devised by Gill [2.4] and has proved suitable for school use. More sophisticated battery-operated oxygen meters are now available which make the measurement of oxygen in water much simpler.

Hydrogen-iron concentration (pH)

Water originating from limestone soils is alkaline, with a pH of 7.5 or above. Heathland peaty soils give rise to acid water with a pH of 6.0 and below. The usual method for estimating pH is by using a multiple soil indicator such as that supplied by British Drug Houses. This includes colour charts reading to the nearest 0.2 pH. A sample of water to be tested is drawn up to the first mark in a capillator. Indicator solution is then drawn up to the top mark and the colour compared with the chart. If there is a muddy suspension of particles in the sample to be tested, addition of barium sulphate will cause flocculation of the particles without altering the pH and the clear fluid can then be tested.

Fieldwork

1 Record water temperature in a pond or lake over a period of time, taking readings at the same place and depth on each occasion. Record the surface temperature also at these sites. Plot a graph of these recordings and, if possible, relate the differences of temperature to the distribution of organisms at the sites. Depending on the sample size, mean temperature differences between seasons can be calculated from the data using either an analysis of variance or the '*t*' or '*d*' statistic [2.1].

2 To demonstrate the strength of the meniscus, float a steel sewing needle on the surface of a glass of water. Then add a drop or two of liquid detergent. The needle will sink. Repeat the experiment when a surface-dwelling insect such as a pond skater is introduced to the surface of water in a small glass vessel. Note the layer of air surrounding the insect as it sinks beneath the surface. The insect should be transferred to fresh water as soon as possible.

References

2.1 Bishop, O. (1981) *Statistics for Biology,* 3rd edn. Longmans.
2.2 Dowdeswell, W. H. and Humby, S. R. (1953) 'A photo-voltaic light meter for school use', *School Sci. Rev.,* **35,** *No. 125,* 64–70.
2.3 Dowdeswell, W. H. (1984) *Ecology: Principles and Practice,* Heinemann.
2.4 Gill, B. F. (1977) 'A plastic syringe method for measuring dissolved oxygen in the field or laboratory', *School Sci. Rev.,* **55,** *No. 214,* 458–60.
2.5 Knight, C. B. (1965) *Basic Concepts of Ecology,* Macmillan.
2.6 Mortimer, C. H. (1956) 'The oxygen content of air-saturated freshwaters, and aids in calculating percentage saturation'. *Mitt. int. Ver. Limnol.,* **6.**
2.7 Schwoerbel, J. (1970) *Methods of Hydrobiology. Freshwater Biology,* Pergamon Press.

Table of the solubility of oxygen in chloride-free water at various temperatures when exposed to water-saturated air at a total pressure of 760 mmHg and partial pressure of oxygen at 160 mmHg. (Dry air is assumed to contain 20·9 per cent oxygen.)

°C	ppm	°C	ppm	°C	ppm
0	14·62	17	9·74	34	7·2
1	14·23	18	9·54	35	7·1
2	13·84	19	9·35	36	7·0
3	13·48	20	9·17	37	6·9
4	13·13	21	8·99	38	6·8
5	12·80	22	8·83	39	6·7
6	12·48	23	8·68	40	6·6
7	12·17	24	8·53	41	6·5
8	11·87	25	8·38	42	6·4
9	11·59	26	8·22	43	6·3
10	11·33	27	8·07	44	6·2
11	11·00	28	7·92	45	6·1
12	10·83	29	7·77	46	6·0
13	10·60	30	7·70	47	5·9
14	10·37	31	7·50	48	5·8
15	10·15	32	7·40	49	5·7
16	9·95	33	7·30	50	5·6

Appendix 2B

The Winkler method of estimating dissolved oxygen

The method depends upon the fact that when a manganese (II) salt is added to water containing oxygen in alkaline conditions a proportion of manganese (II) is oxidized to manganese (III). Upon acidification in the presence of potassium iodide, iodine is liberated in an amount equivalent to the dissolved oxygen present. The free iodine is then estimated by titration against a standard thiosulphate solution, using starch as an indicator.

Reagents

A — 40 per cent manganese (11) salt solution.
B — 33 g sodium hydroxide and 10 g potassium iodide dissolved in 100 cm^3 distilled water.
C — N/80 sodium thiosulphate (VI) solution.
D — Concentrated hydrochloric or ortho-phosphoric acid.
E — Starch solution, freshly prepared for each series of titrations.

Procedure

1 The sample of water to be tested, which has not come into contact with air, is contained in a stoppered bottle of 70 cm^3 capacity. (For method of obtaining such a sample see Dowdeswell [2.3]).
2 Remove stopper and, using a pipette, run in 5 cm^3 of A and 3.1 cm^3 of B to the bottom of the bottle. (For volumes greater or less than 70 cm^3 calculation of the amounts of A and B to be added must be made). Ensuring that no air bubbles are included, replace stopper and shake. Leave to stand for 5 minutes when a brown precipitate of manganese (III) will appear and oxygen in the sample is now fixed. This operation is best carried out in the field.
3 Add about 2 cm^3 of D, replace stopper and shake. Iodine will be liberated.
4 Titrate 25 cm^3 of the liquid against solution C to which a drop or two of freshly prepared indicator E has been added. Repeat procedure using a second sample.

Calculation of oxygen content of sample

$$1\ cm^3\ N/80\ Na_2S_2O_3 \approx 0.1\ mg\ O_2$$

The amount of oxygen l^{-1} can be calculated as follows:

$$\text{mg oxygen } l^{-1} = \frac{V \times 0.1 \times 1000}{v}$$

Where V= volume of thiosulphate (VI) used and v = volume of sample.

$$1\ cm^3\ N/80\ Na_2S_2O_3 \approx 0.0001\ g\ oxygen$$

therefore $$1\ cm^3\ N/80\ Na_2S_2O^3 \approx 0.0001 \times 22\,400\ cm^3\ oxygen$$

therefore $$\frac{V \times 0.0001 \times 22400 \times 1000}{32 \times v} = \frac{V \times 70}{v}$$

3 Plankton and the freshwater community

Plankton can be defined as very small organisms which live suspended in the water. Their bodies may be denser than water but nevertheless they are kept in suspension, moved by convection or wind-induced currents. **Phytoplankton** are microscopic green plants, comprising a large number of species of algae and photosynthetic bacteria. Microscopic animals that drift with the current form the freshwater **zooplankton**. These are mostly small crustaceans and rotifers. All plankton form an important part of food chains.

The nature of the plankton community

Every cubic centimetre of lake or pond water contains thousands of planktonic organisms. Some are microscopic plants and bacteria, others include fungi and microscopic animals which feed on living or dead matter or upon soluble organic material. All are capable, under favourable conditions, of rapid multiplication, producing excretions and secretions. When they die their bodies add large amounts of organic material to be degraded by other organisms and recycled.

In describing the planktonic community the question of size arises. The range of variation is great, from the host of bacteria, some of which are only 0.1 μm in diameter, to some of the zooplankton which are 50 μm or more. Species of phytoplankton, especially when they occur as colonies, may be 100 μm in diameter or greater and visible to the naked eye; while many of the zooplankton species, such as the crustaceans *Daphnia*, *Cyclops* and *Diaptomus* (Fig. 3.3) are larger still.

Freshwater phytoplankton

Taxonomically, as well as in size, the phytoplankton form a very diverse group. Some species of bacteria which contain chlorophyll are capable of photosynthesis. They are especially abundant in conditions of low oxygen levels, such as the deeper parts of fertile lakes, and in smaller bodies of water where there is an abundance of fallen leaves. These photosynthetic bacteria may be extremely important in the recycling of sulphur and carbon.

By definition, phytoplankton are microscopic plants which are suspended in the water. Nevertheless, it is a common misconception that phytoplankton float, having similar density to water. This is not always so, since some possess gas vesicles which give them buoyancy. The degree of buoyancy is dependent upon the light intensity which, in turn, determines the amount of gas. Many of these small plants and animals have gelatinous sheaths containing water to reduce their density. Some blue-green algae may float at particular depths which favour their growth processes. Others, such as *Botryococcus* (Fig. 3.1 (a)), store a large amount of oil which provides positive buoyancy.

Diatoms (Fig. 3.1 (e)) have cell walls composed of silica (SiO_2) and because they have a specific gravity of as much as 0.06 in excess of that of water they will sink unless transported by wind-generated currents. Many algae, e.g. *Euglena* or *Chlamydomonas* (Fig. 3.1 (b),(c)), possess flagella which, by their beating action, help to counteract a tendency to sink. Others have developed spines and other forms of projection which increase their resistance to the water and therefore help to delay sinking. However, by no means all the diverse species of phytoplankton possess devices to prevent sinking, and it is quite possible that a slow

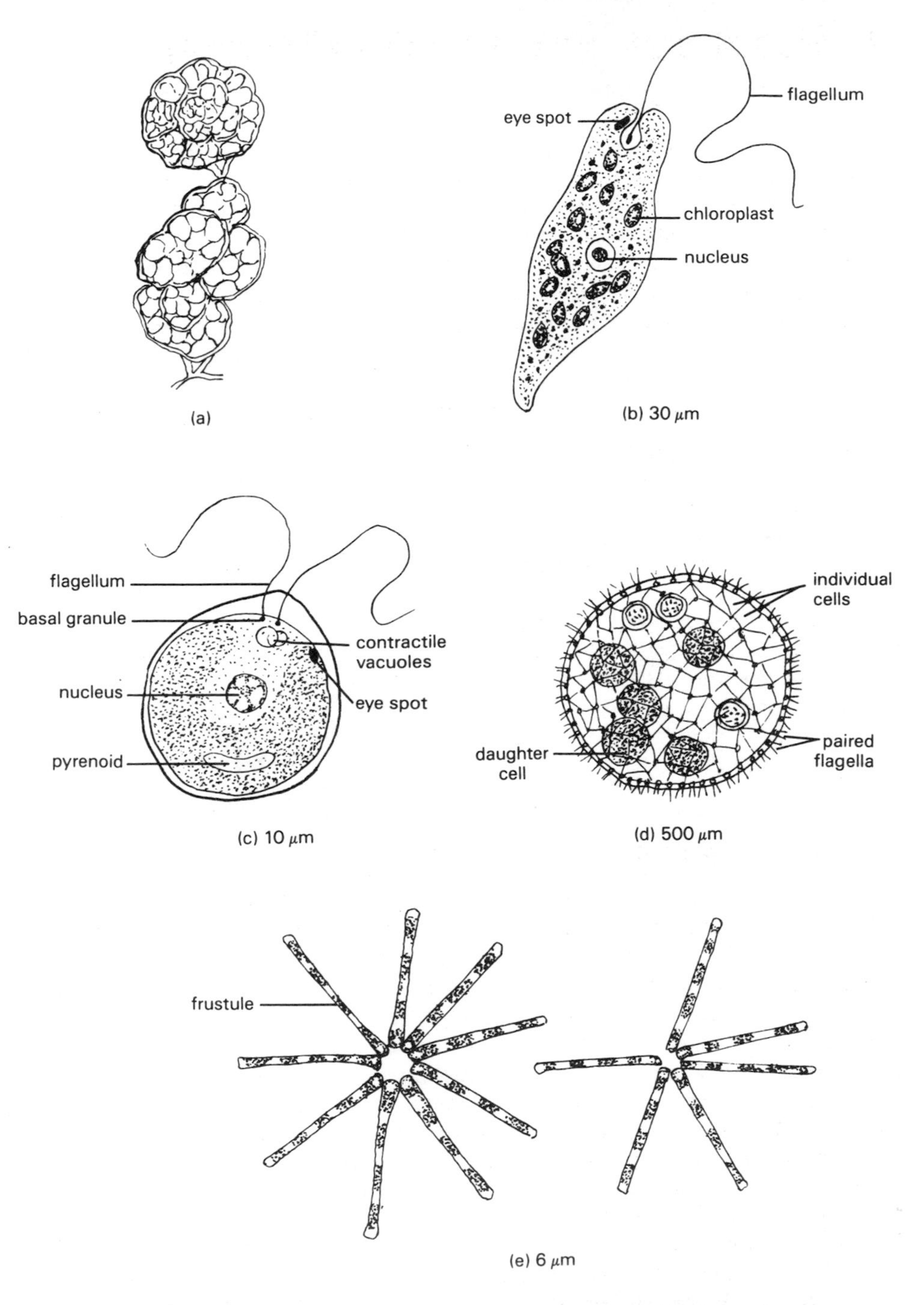

Fig. 3.1 Some examples of phytoplankton. (a) *Botryococcus braunii*, a yellow/green alga, forms strings of indefinitely-shaped structures containing oil globules. (b) *Euglena* and (c) *Chlamydomonas* possess flagella which, by their beating action, provide positive buoyancy. (d) *Volvox* is a colonial alga composed of hundreds of individuals. A rotifer, *Ascomorpha volvocicola*, lives parasitically within *Volvox*, feeding on the green cells. (e) *Asterionella* is found in huge numbers floating free. Between six and nine frustules, each a separate diatom, compose a star-shaped structure

downward progression can be a positive advantage, enabling the organisms to absorb nutrients by diffusion at the cell surface.

Most planktonic algae tend to form cell masses rather than existing singly. This increases their surface area relative to volume, thereby providing friction to reduce the rate at which they sink. Some of these colonial forms assume quite large spherical structures as much as 0.2 mm in diameter such as those of *Volvox* (Fig. 3.1 (d)), which in early summer can easily be seen with the naked eye as green spheres near the surface of the water. Such large multicellular globes are less readily preyed upon by filter-feeding zooplankton.

Phytoplankton are vulnerable to grazing herbivores and can also be swept by flooding from their habitat or even lost by sinking. Such hazardous conditions favour rapid methods of reproduction to replace losses and most species have abandoned sexual reproduction. The simple division of one cell into two, which can occur in a matter of hours, is the answer, while fertilization involves a union of mobile sperm with relatively immobile egg cells in conditions where such cells may be widely separated. The asexual form of cell division means the limitation of genetic variation, which can be a relatively unimportant factor in an environment which calls for little variation.

Populations of small phytoplankton, up to 10 μm in size, exist in numbers as large as 10^6 ml^{-1}. Even so, individuals in such large populations, occurring under favourable conditions of light and nutrients, are fairly widely spaced which may explain why parasitism by protozoans and fungi is relatively rare. Fluctuation in population sizes is due to sudden changes in the temperature, the amount of nutrient present and the light intensity.

Freshwater zooplankton

Protozoa, rotifers, cladocerans, and copepods are the chief groups contributing to freshwater zooplankton, as well as a few coelenterates which are predators of other zooplankton.

Among the common species of protozoa are *Arcella* and *Difflugia* (Fig. 3.2 (a,b)). Both can extend pseudopodia which assist in preventing them from sinking.

Rotifers, or wheel animalcules, have a complex internal structure with reproductive and excretory systems. Although multicellular, the cells have no distinct cell walls. A circlet of cilia surrounds the mouth region and these beat rhythmically to give a seemingly rotational motion. They are all suspension feeders, gathering particles by means of cilia, up to 20 μm in size.

Rotifers rarely reach a length of more than 2 mm, the males being often much smaller than the females. A study of living rotifers reveals an endless variety of forms, some spherical and free-swimming, others flattened or even worm-like. However, although most are free-swimming in their early stages, after a short time some species cease to move freely and become permanently fixed or sessile, and can even adopt a creeping habit. Figure 3.2 (c,d) shows two common species of rotifer. Under a microscope, using dark-ground illumination, rotifers can easily be seen, especially if their movements are restricted by the addition of a drop or two of glycerol to slow down their movements.

Both cladocerans and copepods are crustaceans and members of both groups contribute large numbers to the zooplankton.

The cladocerans include the well-known genus of water fleas, *Daphnia* (Fig. 3.3 (a)). Daphnia possess five pairs of limbs enclosed by a bivalve shell, or carapace, within which the limbs constantly beat, drawing a current of water through the carapace. Movement is brought about by the flicking action of the long, branched second antennae. These have a fringe of closely spaced setae which filter fragments of food from the water and convey them to the mouth. The large eye is very obvious and the internal organs are clearly visible through the transparent carapace.

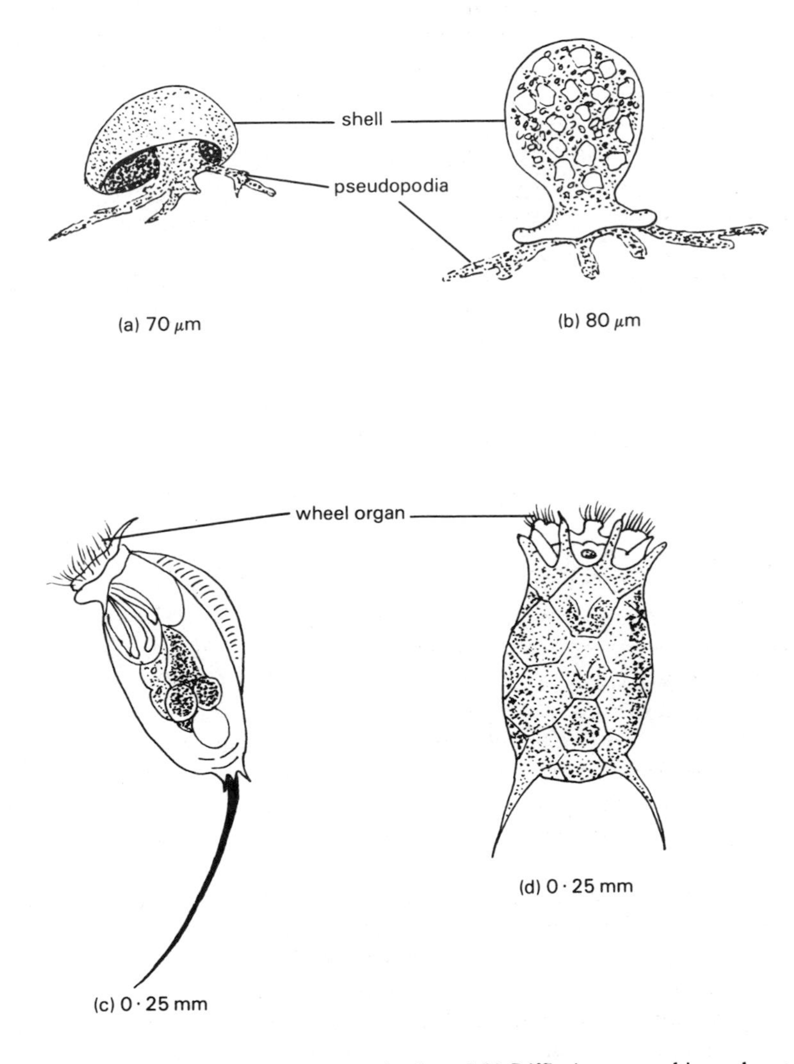

Fig. 3.2 Some examples of zooplankton (a) *Arcella* and (b) *Difflugia* are two rhizopod protozoans found on the surface of mud rich in organic matter. *Arcella* secretes a smooth shell of chitin. In *Difflugia* the shell is composed of sand grains attached to a secreted layer. (c) *Trichocerca bicristata* and (d) *Keratella quadrata* are two free-swimming rotifers, their posterior spines assisting buoyancy

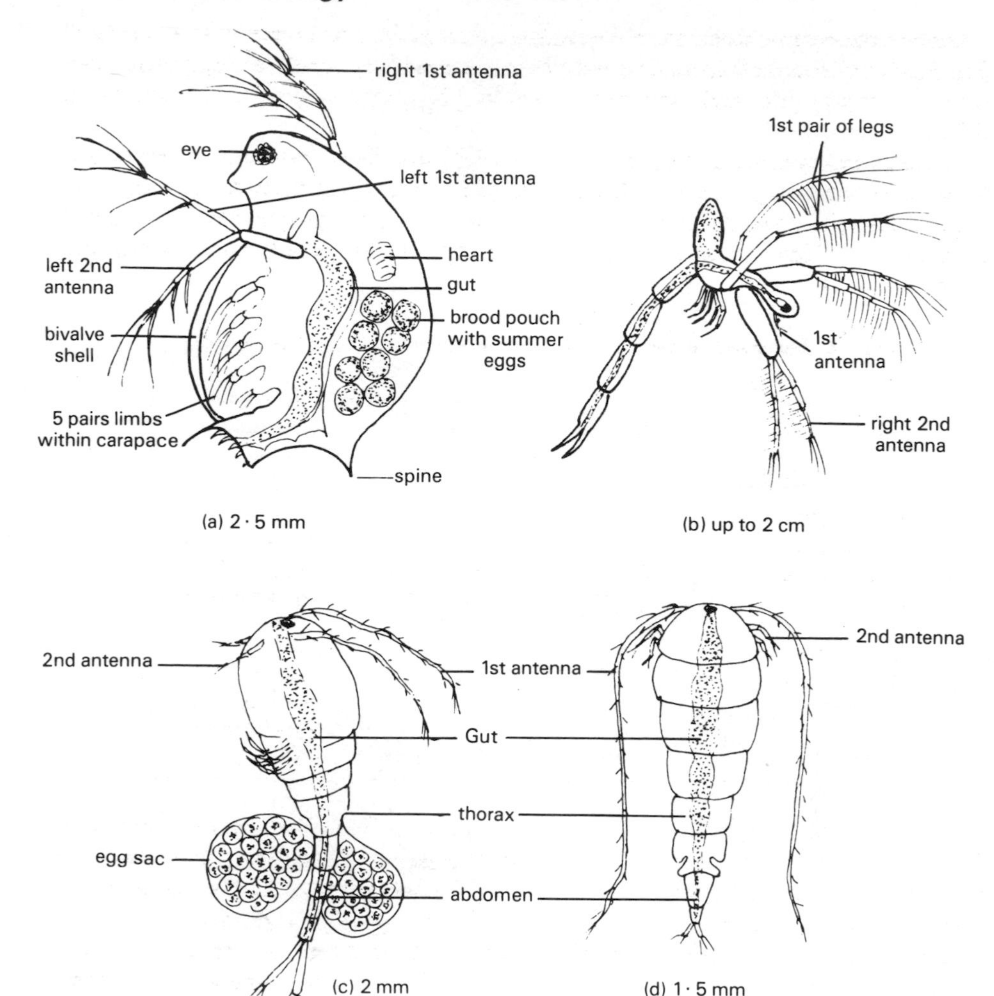

Fig. 3.3 Planktonic crustaceans. (a) *Daphnia pulex* is a filter-feeding cladoceran. The 2nd antennae, which are also the main swimming organs, waft particles of food towards the mouth. The other limbs keep a constant respiratory current flowing through the transparent bivalve shell. In summer the females produce a number of parthenogenetic eggs within the brood chamber, which hatch in a few days. In winter two resistant eggs are produced and fertilized by smaller males. At the next moult, these eggs acquire a stout covering, the ephippium, which floats to the surface for dispersal by wind or birds. (b) *Leptodora kindtii*, a carnivorous cladoceran, is completely transparent. The long 2nd antennae propel the animal through the water and also serve to catch prey, usually other smaller crustaceans. (c) *Cyclops* sp. Antennae are half the body length. The female bears two egg sacs. (d) *Diaptomus* sp. Antennae as long as body. Female bears one egg sac. Both (c) and (d) are copepods, among which some are filter feeders and others carnivores, themselves forming the food of fish. Summer eggs develop quickly; in times of drought, resting eggs are produced which can be dispersed great distances by wind or birds

Another cladoceran, common in the surface waters of lakes and reservoirs, is *Leptodora* (Fig. 3.3 (b)). Unlike the filter-feeding water fleas, it is carnivorous and uses its raptorial limbs to capture its prey. Although often nearly 2 cm long *Leptodora*, being quite transparent, is difficult to see.

Copepod crustacea are usually larger than cladocerans. They may be filter-feeders like *Diaptomus* or predatory like *Cyclops* (Fig. 3.3 (c,d)). The size of prey depends upon their own size and ability to grasp the prey. This can be a variety of small zooplankton or masses of phytoplankton. Table 3.1 compares some of the important features in three zooplankton groups.

Table 3.1 Features of the three main zooplankton groups compared (modified from Allan, 1976)

Features	*Rotifera*	*Cladocera*	*Copepoda*
Generation time (days)	1·25–7	5·5–24	7–32
Adult body length (mm)	0·2–0·6	0·3–3·0	0·5–5·0
Food size (μm)	1–20	1–50	5–500
Method of feeding	Suspension feeding	Filter-feeding (raptorial in carnivores)	Filter or raptorial feeding
Filtering rate	Very low	High	Low
Susceptibility to predators:			
invertebrate	High	Moderate	Moderate (adults) High (juveniles)
vertebrate	Very low	High	Low

A rare but interesting species is *Craspedacusta sowerbyi* (Plate 3.1), the only species of freshwater medusa in Great Britain. It feeds on small planktonic organisms in much the same way as the large marine jellyfish. One specimen of *C. sowerbyi* turned up in a plankton net in the Exeter Canal, Devon, during hot weather in September, 1976. Individuals, sometimes in large numbers, have been recorded in other localities during the unusually warm summers of 1928 and 1947. Their occurrence may, therefore, be connected with above-average water temperatures.

Distribution of plankton

Certain factors favour rapid growth and reproduction of plankton. Phytoplankton, because of their photosynthetic activities, require light but a rise in temperature increases their growth and multiplication, which is also correlated with the availability of suitable nutrients. Turbidity cuts down the amount of light penetrating the water and therefore the rate of photosynthesis; it may also interfere with the feeding mechanisms of zooplankton.

Both phytoplankton and zooplankton often show a difference in body size at different times of the year. *Asterionella* (Fig. 3.1 (e)), often forms large colonies in summer while some of the cladocerans, such as *Bosmina*, grow a hump on the back in summer and a longer proboscis, both of which are lacking in winter forms of the same species (Fig. 3.4). These summer forms have a greater surface area which offers more resistance to the water and therefore they sink more slowly.

One of the most interesting activities of planktonic crustacea, such as *Daphnia*, is their ability to perform vertical migrations. Such a phenomenon has been observed and recorded in a number of freshwater lakes in different parts of the world. At midday the organisms are

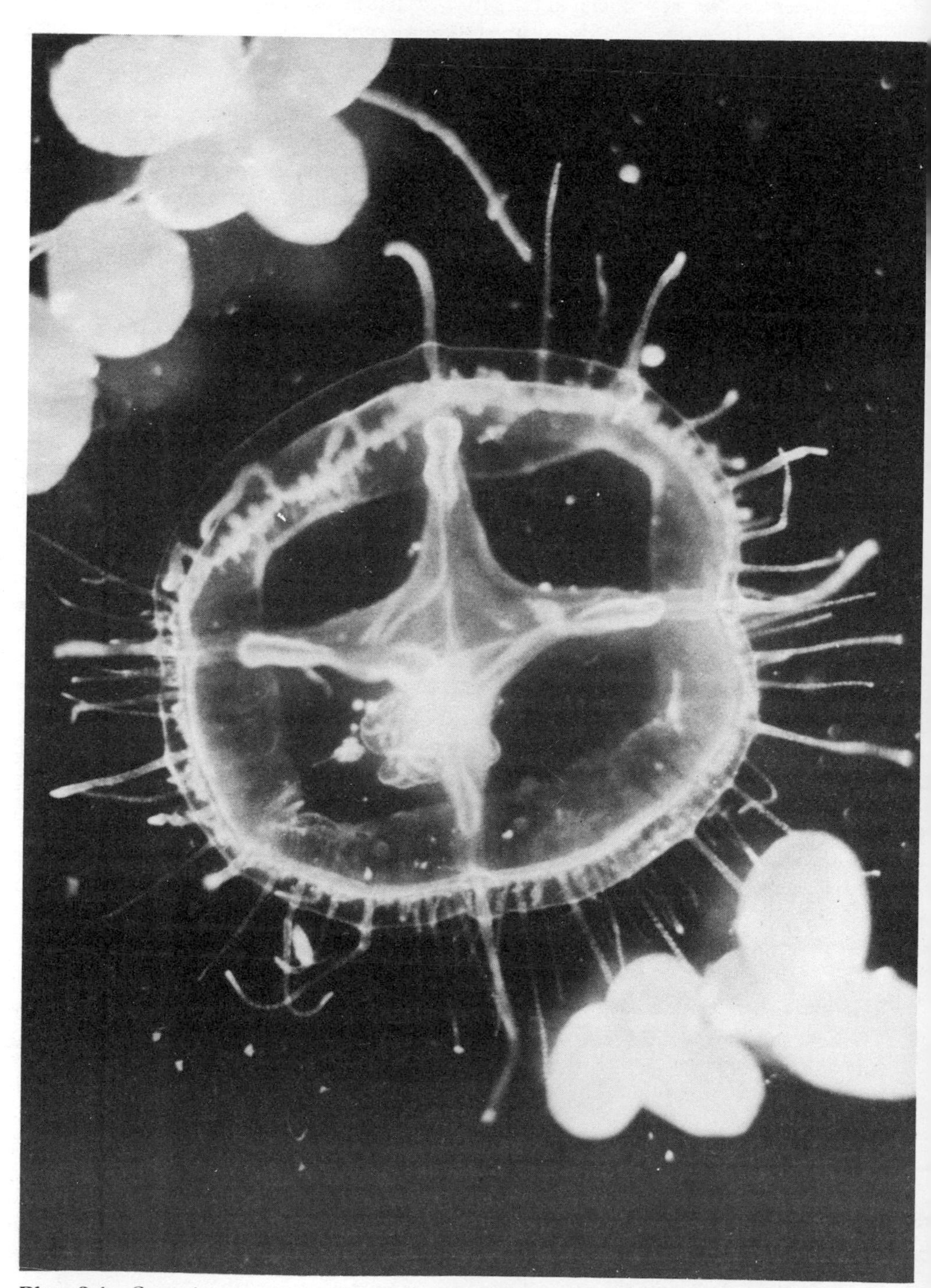

Plate 3.1 *Craspedacusta sowerbyi*, the only species of freshwater medusa in Great Britain. An indication of its size is given by the floating fronds of duckweed (Photo: D. Nichols).

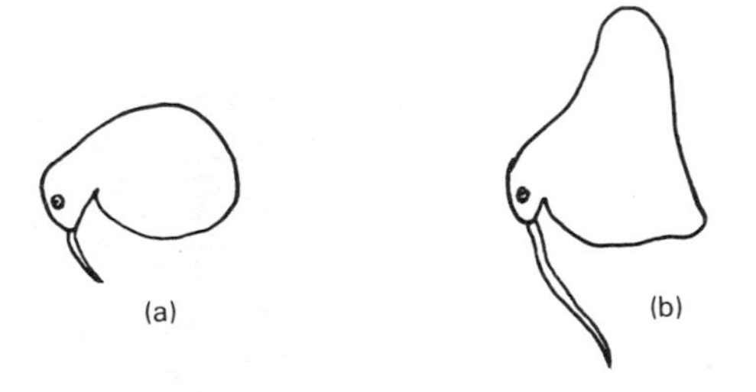

Fig. 3.4 *Bosmina* (Cladocera): (a) winter form, (b) summer form

most abundant at some depth below the surface; while as the daylight fades there is an upward migration (Fig. 3.5). This vertical migration seems to be correlated with the depth to which rays of light at the blue end of the spectrum penetrate.

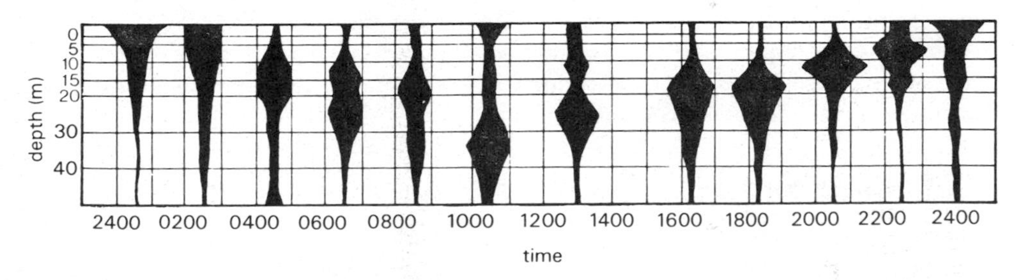

Fig. 3.5 The daily migration of *Cyclops strenuus* in Lake Windermere. The catch at each depth has been converted to percentage of total catch (after Ullyott, 1939)

Running water also contains planktonic organisms but their numbers will depend upon the rate of flow, as well as upon the factors already described for still waters. Eventually both phytoplankton and zooplankton in running water will be lost by being swept out by the current.

The contribution of plankton to the dynamics of the ecosystem

As in other freshwater communities, the plankton contribute to the economics of the ecosystems. Bearing in mind their small individual sizes, their gross contribution to energy production is far greater than one might suppose. Many chemical changes are going on simultaneously and at a very rapid rate. Phosphates and carbon, as well as other substances, are ingested by some species, which are then eaten by others. In this way such substances can be transferred from one compartment of the plankton to another in a matter of minutes. So there is a constant interchange which makes measurement of production relatively difficult.

Photosynthesis in phytoplankton can be expressed as the total amount of oxygen evolved in a given time. This is usually estimated by the light and dark bottle method (Appendix 3). In theory, estimation of the amount of oxygen evolved will give a measurement of **gross** photosynthesis since this will include a correction for the amount of oxygen used in respiration.

Production by phytoplankton can be limited by the absence of substances required for growth. Grazing by zooplankton can reduce their numbers. Even the secretion of organic compounds by the phytoplankton themselves will reduce their metabolic processes and therefore their growth. Other losses can be due to the organisms sinking into the sediment and being washed away. All these processes imply a large daily conversion of phytoplankton to detritus.

Phytoplankton are relatively evenly distributed in fresh water and investigation of their production is therefore quite easy. But the zooplankton community, comprising a large number of species, distributed unevenly, shows marked differences in diet and feeding habits, varying both between individuals of different species as well as between different ages of the same species. Estimating their productivity, therefore, poses considerable problems.

Methods of estimating production in zooplankton are further complicated because of predation by other non-planktonic organisms.

So we see that the broad picture of the economics of plankton is one of rapid transfer of essential nutrients between the water and the living organisms. The essential nutrients such as phosphorus and nitrogen, occurring as soluble salts, are assimilated and combined by the phytoplankton to become available for the energy requirements of the zooplankton or else they are lost by sedimentation to become available to benthic animals.

Finally, while the cycling of substances is a matter of hourly changes within the community, seasonal periodicity involving changes in temperature, nutrients, and light intensity has more long-term effects.

Collecting and counting plankton

For plankton a special net is required. The net consists of a stout wire ring to which is attached tough material of mesh size 0.075 mm or for larger phytoplankton and zooplankton, 0.3 mm. As the net is towed or drawn through the water it filters off the plankton which wash down the net and collect in the container (Fig. 3.6 (a)). For the tow-net, cord must be attached of a suitable length to tow the net along the surface behind a boat. If sampling is to be done at a certain depth beneath the surface, a weight will have to be attached and the length of cord and speed of towing must be adjusted in order to keep the net at a constant depth.

For a hand-sweep net a pole, preferably in several sections to allow for sampling at different depths or areas, is attached. A wing-nut and screw attachment of net to pole is the strongest and enables quick removal of the net. The collecting chamber can be a tube of toughened glass or perpex (Fig. 3.6 (b)).

Both nets should be strengthened down the sides and round the mouth with canvas. After use they should be washed and dried to prevent rotting.

A **cavity chamber** for counting plankton can easily be constructed in the laboratory using a glass microscope slide on which 10 mm squares are etched with a diamond and then rubbed over with a wax pencil to delineate the lines. Strips of microscope slide glass are cemented onto the base slide around the squared area with a resin glue such as 'Araldite', to form the sides of the chamber.

Fieldwork

Using one of the nets described above, sample the plankton in a lake or pond to find out if the density of plankton varies at different depths, different times of day or at different seasons. Sample size should be standardized by towing the net over a fixed distance or, if using a hand net, by the number of sweeps.

Plankton die quickly, especially with a rise of temperature when a sample is removed from the water. The sample can either be stored in a thermos flask or 'fixed' by the addition of a small quantity of 5 per cent formalin. Numbers can be counted and identified by using a

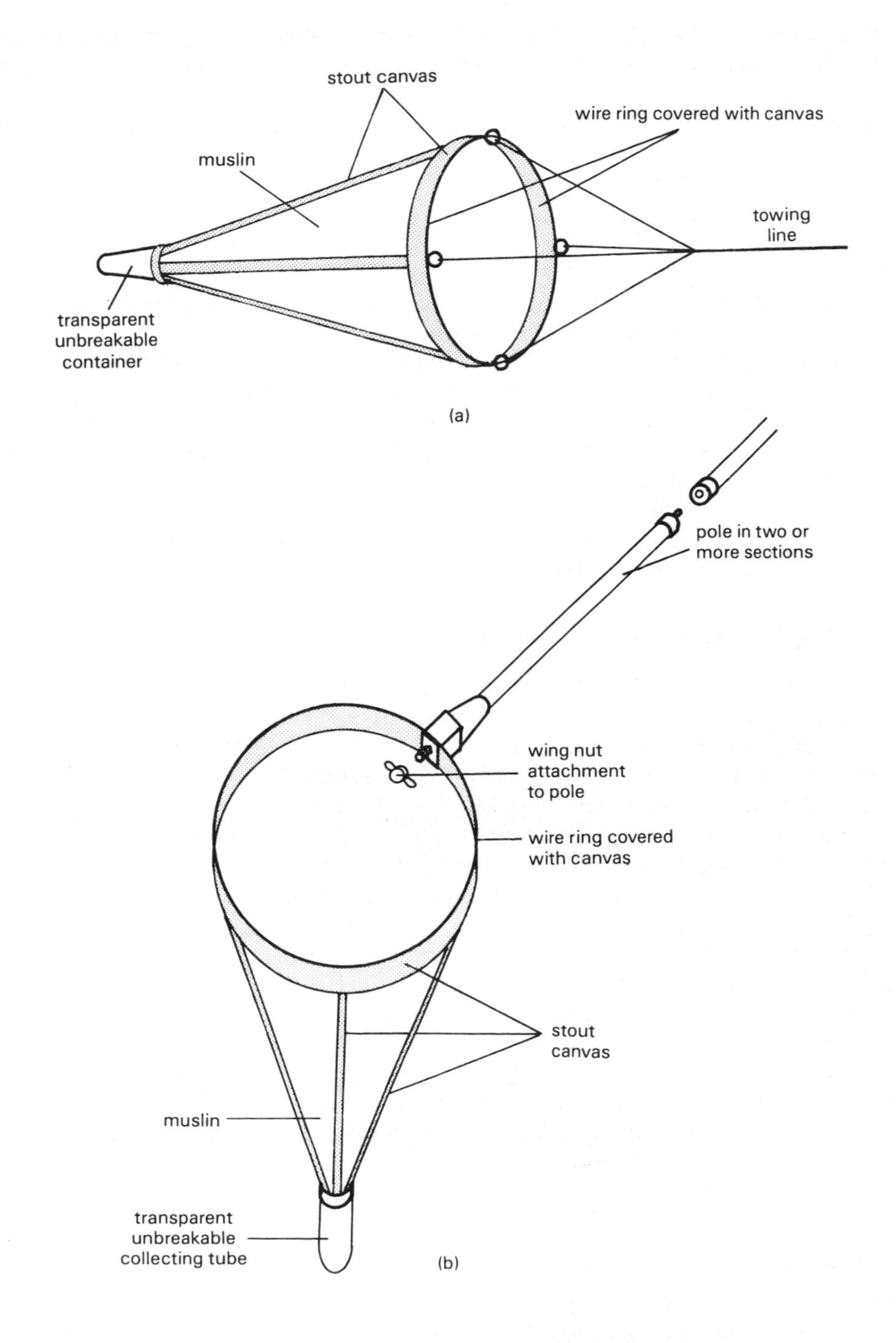

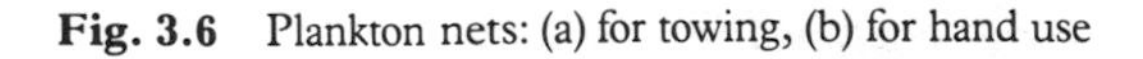

Fig. 3.6 Plankton nets: (a) for towing, (b) for hand use

cavity chamber. After shaking the sample to ensure randomization, transfer a part by means of a micro-pipette to the chamber and, being careful not to include air bubbles, lower a long cover slip on top and examine under the low power of a binocular microscope. The following points should be taken into account:

(i) The density of plankton may be too great for an accurate count and dilution will then be necessary.
(ii) Recording the number of each species in each of the 10 mm squares will prove difficult unless the number of species recorded is limited to five or less.
(iii) Since plankton distribution is not uniform, a number of counts should be made using different samples from the same locality.
(iv) In the first instance, it may be advisable to confine attention to the larger and more easily identifiable organisms, using books mentioned in the 'Identification' section of the Bibliography.

The occurrence of a species is then calculated as a percentage of the total number of counts made of that species. This is the **percentage frequency**.

From these results compare contrasting batches by estimating the **standard deviation** of the means of samples and the **standard error** (see the references at the end of this chapter for methods of calculation).

Is there any relationship between the results obtained at different depths, times of day, or at different seasons?

References

3.1 Allan, J.D. (1976) Life history patterns in 300 plankton, *Am. Nat.*, **110**, 165–80
3.2 Bishop, C.K. (1981) *Statistics for Biology*, 3rd edn, Longmans
3.3 Campbell, R.C. (1974) *Statistical Methods in Biology*, 2nd edn, Cambridge
3.4 Dowdeswell, W.H. (1984) *Ecology: Principles and Practice*, Heinemann

Appendix 3

Estimating the photosynthetic production of phytoplankton using the light and dark bottle method

A sample of the population of phytoplankton is enclosed in a clear glass bottle (light bottle) completely full of bubble-free water and left in light conditions for a known period of time. The amount of oxygen released is then measured by the Winkler method (Appendix 2B). Under these conditions, oxygen is evolved during the process of photosynthesis but it is also absorbed by respiration of the phytoplankton and any small animals or bacteria present in the sample. To compensate for this, a replicate sample, shielded from light by enclosure in a dark bottle (a bottle painted with black paint to occlude all light), is left for the same period of incubation allowed for the light bottle and the oxygen content is then measured. The difference between the values obtained for the dark and light bottles gives a measure of the amount of oxygen which has been evolved over that period. This is directly proportional to the amount of carbon dioxide incorporated during photosynthesis.

4 Still waters of lakes and ponds

Freshwater ecosystems are usually divided into two categories: lentic (or still) bodies of water and lotic (or running) water. In this chapter we are concerned with the former which range from small, shallow, temporary pools to large lakes. Within this range there are other important variants which we must consider such as altitude, the geology of the catchment area and, above all, the depth of the water. Since most light energy is absorbed in the uppermost 3 metres (p 4), the actual volume of water receiving solar radiation will depend on the area of the lake. This, in turn, will affect productivity because, as we have already seen (p. 11), the photosynthetic activities of the plants are governed by the overall amount of light reaching them.

Origins of lakes

Lakes originate in various ways as a result of movements of the Earth's crust: for instance, by man's activities in mining and quarrying, by volcanic activity, by the action of glaciers, by landslides, and occasionally by the impact and explosion of meteorites. Animals and plants can cause damming by building up organic material, thereby blocking valleys. Man-made lakes and reservoirs are formed by constructing dams across rivers flowing through steep-sided valleys or by excavation (e.g. in mining or quarrying).

Once a lake basin has been formed it can change over a period of time and even eventually be destroyed by a variety of filling-in processes. Thus wearing away or erosion of the rim of the lake can occur due to the scouring action of the water. Again, streams and rivers entering the lake bring in sediments, while the activities of organisms living in the lake will add detritus to these sediments. The final process of filling in is accelerated by the growth of marginal plants upon the sediments and gradual encroachment towards the centre of the basin. These processes are more obvious, and can take place more quickly, in a pond than in a large lake. Such a transition from water to land in known as a **hydrosere** (Plate 4.1).

It is difficult to make a precise distinction between what we mean by a pond and a lake because there can be so many transitional steps. Generally speaking we can say that a pond is a body of water which is shallow enough for light to reach the bottom all over, which does not show a complete temperature stratification and whose shores are not eroded by wave action. A lake, on the other hand, is a body of water in which thermal stratification does take place, has depths to which light rays cannot penetrate sufficiently for photosynthesis to take place and whose shores are often eroded by wave action.

The nature of a lake depends upon a number of factors such as its basin shape, its substrate, and the amount of nutrients present.

Depth and stratification

For small bodies of water, comparison of average temperature measured at a depth of 20 cm below the surface with the weekly average air temperature, based on the daily maxima and minima, shows close correlation. However, in lakes with a large volume of water and a greater depth the situation is different. Following the coldest part of the winter, there may be a period when the lake water has a temperature approaching 4°C from top to bottom. As the air temperature rises in the spring, the upper layers are warmed, becoming less dense. But since the Sun's rays are absorbed by water the deeper parts remain cold and dense, although

Plate 4.1 Hydrosere. Note the encroachment of reedswamp and willow at the edge of the peat pool (Somerset Wetlands)

a stormy period can cause the warm upper and cold lower waters to mix. As the summer progresses the two layers become established, with such a large temperature difference between them that they remain separated. These conditions, which are termed **stratification**, can prevail in a lake such as Windermere. Figure 4.1 shows that by midsummer the temperature of the lower zone of the lake, the **hypolimnion**, varies only about 2°C over a depth of 15 and 60 m and that of the upper zone, the **epilimnion**, varies only 1°C from the surface down to about 8 m. Between the epilimnion and hypolimnion there is a zone of rapid change, the **thermocline**, in which the temperature falls rapidly. During the colder autumn weather, heat loss from the surface causes a cooling of the epilimnion and when the temperature reaches that of the hypolimnion, their waters start to mix again (Fig. 4.2).

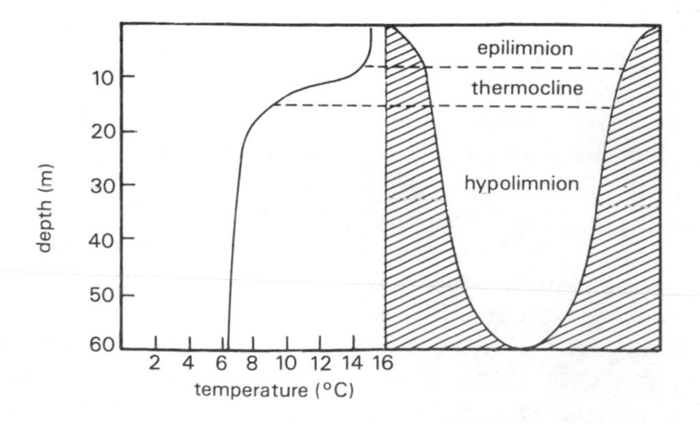

Fig. 4.1 Temperatures taken on a July day in Lake Windermere at different depths (redrawn from Macan and Worthington, 1951)

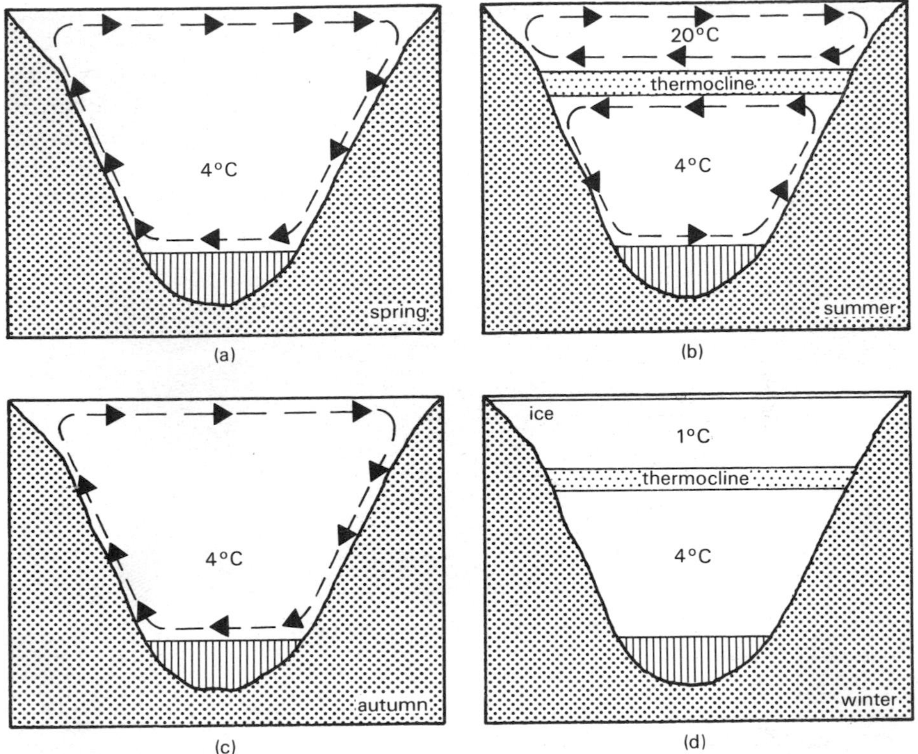

Fig. 4.2 Hypothetical annual thermal cycle in a temperate lake. (a) Water temperature uniform from surface to bottom. (b) Water temperature stratified as surface temperature rises causing thermocline. (c) Conduction of heat and turbulence causes heat to be transferred from epilimnion to hypolimnion until a uniform temperature is reached from bottom to surface. (d) Inverse stratification takes place as water reaches 4°C, rises and freezes, while warmer water remains near the bottom and a thermocline is again established

In a shallow pond temperature fluctuations can be very great, both diurnally and seasonally, and these variations can have a profound affect on the organisms living in the pond.

Light and sediments

As we have already seen in Chapter 2 (p. 11), any material, either floating or in suspension, will cause an obstruction to the penetration of light, thereby decreasing photosynthetic activity of the plants. But apart from its importance in this respect it must be realized that all material, whether of living or non-living origin and whether suspended or in sedimented deposits, represents a reservoir of nutrients available to the organisms living in a lake or pond.

Detrital material in a lake or pond is brought in by inflowing tributaries or by wind. Much of this matter consists of fallen leaves and other vegetation as well as peaty material from upland boggy areas. It can also include pollutants introduced by man.

When there is a period of complete circulation of water between surface and bottom, especially in shallow bodies of water, large amounts of sedimented matter on the bottom are stirred up and transferred to other areas or kept in suspension for long periods until once more sedimented (Fig. 4.3).

The dead bodies of organisms in the epilimnion are broken down by bacteria and fungi into simple soluble substances. These may be used at once by plants, the finer particles remaining suspended. The larger fragments fall through the epilimnion and decaying processes continue in the hypolimnion, but at a much slower rate due to the reduced temperature and shortage of oxygen. Eventually, any remaining particles settle as sediment on the bottom.

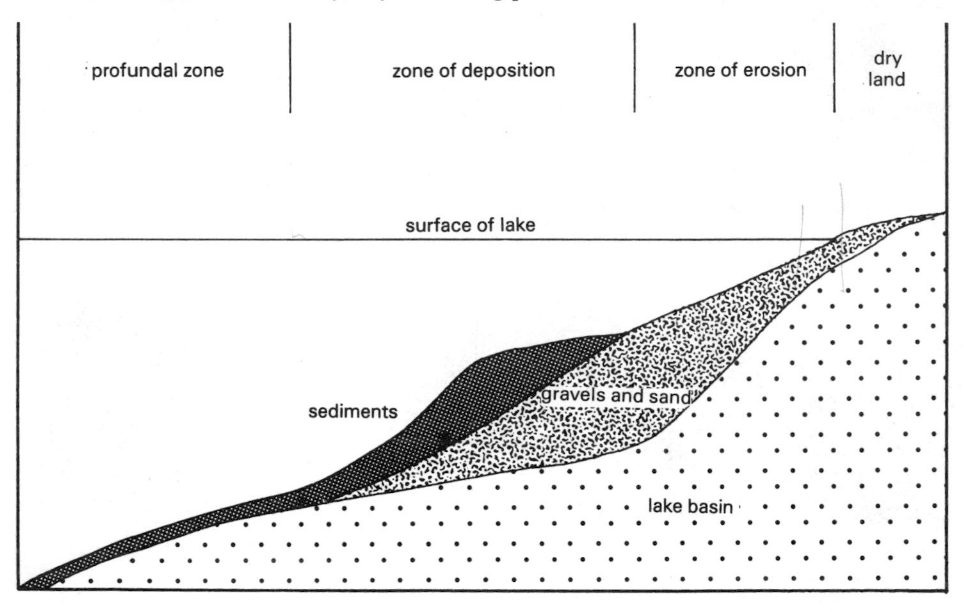

Fig. 4.3 Transverse section through a wave-washed shore showing deposition of sediments

Chemical interrelationships

The amount of oxygen present is far more critical for the organisms in standing waters than in terrestrial habitats because, even when saturated, the amount of oxygen which can be held in water is small compared to that present in air (Fig. 2.4). The rate at which oxygen diffuses through water is also less than in air.

The quantity of oxygen present in a static water ecosystem depends upon:

(i) The area of clear water exposed to the air.
(ii) The circulation of water within the system.
(iii) The amount of oxygen generated and used by the organisms present.

In a shallow, well-weeded pool there may be great fluctuations of oxygen, with super-saturation occurring during the day but with much lower values at night.

Thermal stratification of lake water will have an effect upon the amounts of oxygen present in the different zones. The water of the epilimnion circulates quickly, coming into contact with the air, while water in the hypolimnion circulates slowly and rarely comes into contact with the atmosphere. Fewer green plants are able to exist at these depths and therefore less oxygen is produced as the result of photosynthesis.

The temperature of the hypolimnion is also important and although a lower temperature will mean an increase in the amount of oxygen in solution, the oxygen is constantly being depleted by the respiration of fish and invertebrates.

In lakes rich in nutrients there is a higher standing crop of animals, oxygen will be used up at a greater rate and there is a larger amount of decaying material falling from the upper layers than in lakes poor in nutrients.

In spring, when the temperature of the epilimnion approaches more nearly that of the hypolimnion, the waters will become mixed and may be saturated with oxygen from top to bottom at a concentration which will depend both on temperature and atmospheric pressure and also upon depth. In very deep lakes the temperature of the hypolimnion may remain nearly constant and the water becomes completely deoxygenated. With the coming of the autumn and the cooling of the surface water there will be another period of mixing of the waters of the epilimnion and hypolimnion.

In most lakes the hydrogen ion concentration of the water (pH) and the amount of carbon dioxide in solution are more or less dependent on each other. During daylight the photosynthetic activity of plants uses up dissolved carbon dioxide, thereby increasing the pH, but the situation is reversed at night, reducing the pH once more. The relationship between temperature, oxygen content, pH and amount of dissolved carbon dioxide is well illustrated by the surface recordings made by Tressler *et al* [4.2] as long ago as 1940 in Buckeye Lake, Ohio (Fig. 4.4).

Phosphorus is the scarcest element present in the Earth's crust and yet is essential for growth to algae and higher plants. It is relatively insoluble and is precipitated as salts of iron, calcium, and other metals. The supply of phosphorus is related to that of commoner elements such as calcium and potassium. Therefore the productivity of a body of water is directly correlated with the amount of available phosphorus.

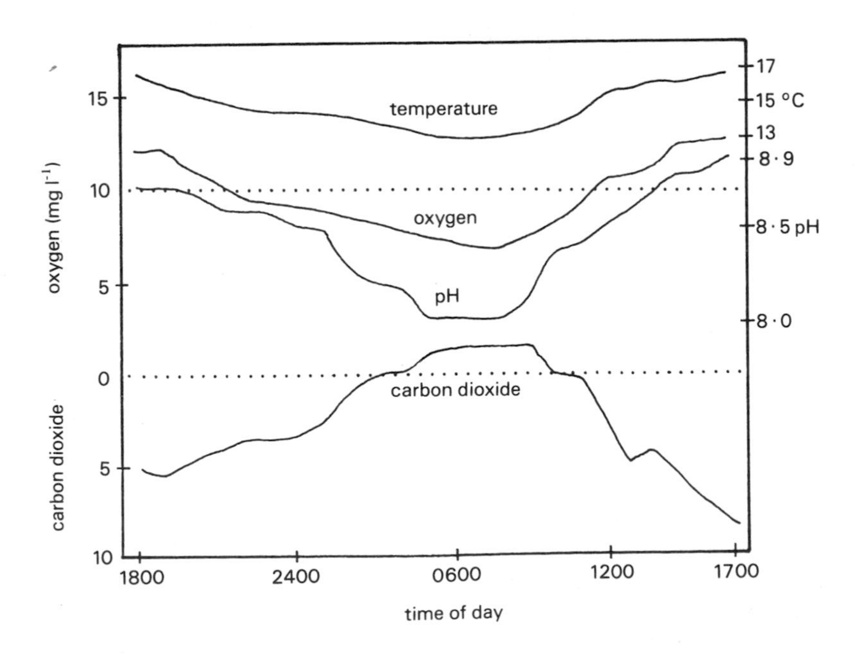

Fig. 4.4 Diurnal fluctuations in temperature, dissolved oxygen, pH and carbon dioxide in the surface waters of Buckeye Lake (Ohio, USA) in summer (after Tressler, Tiffany and Spencer, 1940)

Classification of lakes

It may be useful to end this chapter by a brief reference to the manner in which lakes can be distinguished from one another. Very broadly, there are two kinds: **oligotrophic** lakes are those which are poor in nutrients, deep, and with clear water; **eutrophic** lakes are nutrient rich, usually shallow, and turbid. By increasing the load of nutrients entering a lake, man often causes **eutrophication:** that is, a sudden increase in the algal population thereby also increasing the productivity and causing a reduction of oxygen in deeper regions.

Fieldwork

In the pond or lake you are investigating, can you see any evidence of plant succession from water to land (hydrosere)? If so, you could record such a succession by making a line or belt transect.

A **line transect** involves recording along a straight line, between two poles, the plant species occurring at, say, 0.5 m intervals, or less. This is a quick, although non-quantitative, method of recording the typical species present.

A **belt transect** consists of pegging out two parallel lines, usually 1 m apart, divided across into 1 m squares. The frequency and cover of plants within each square are recorded.

In order to make your record more meaningful in terms of the relationship of the plants to the depth of water, bank, and dry land, a profile can be made by measuring the heights of the plants, and the water and land levels occurring at 0.5 m intervals along one side of the transect. This profile can then be drawn beneath the belt transect chart. Symbols can be used to denote species identified within the transect and recorded in a key. The cover/abundance of each species can be given in brackets after each, using the suggested scale in Fig. 4.5.

From your results, which may not include the same species as those in Fig. 4.5, would you say that they indicate a hydrosere in this part of the pond or lake you are recording? Which plants are typical of open water, mud/marsh, bank, and dry land in the area you are studying?

References

4.1 Macan, T.T. and Worthington, E.B. (1951) *Life in Lakes and Rivers*, Collins

4.2 Tressler, W.L., Tiffany, L.H., and Spencer, W.P. (1940) 'Limnological Studies of Buckeye Lake, Ohio', *Ohio J. Sci.*, 40, 261-290

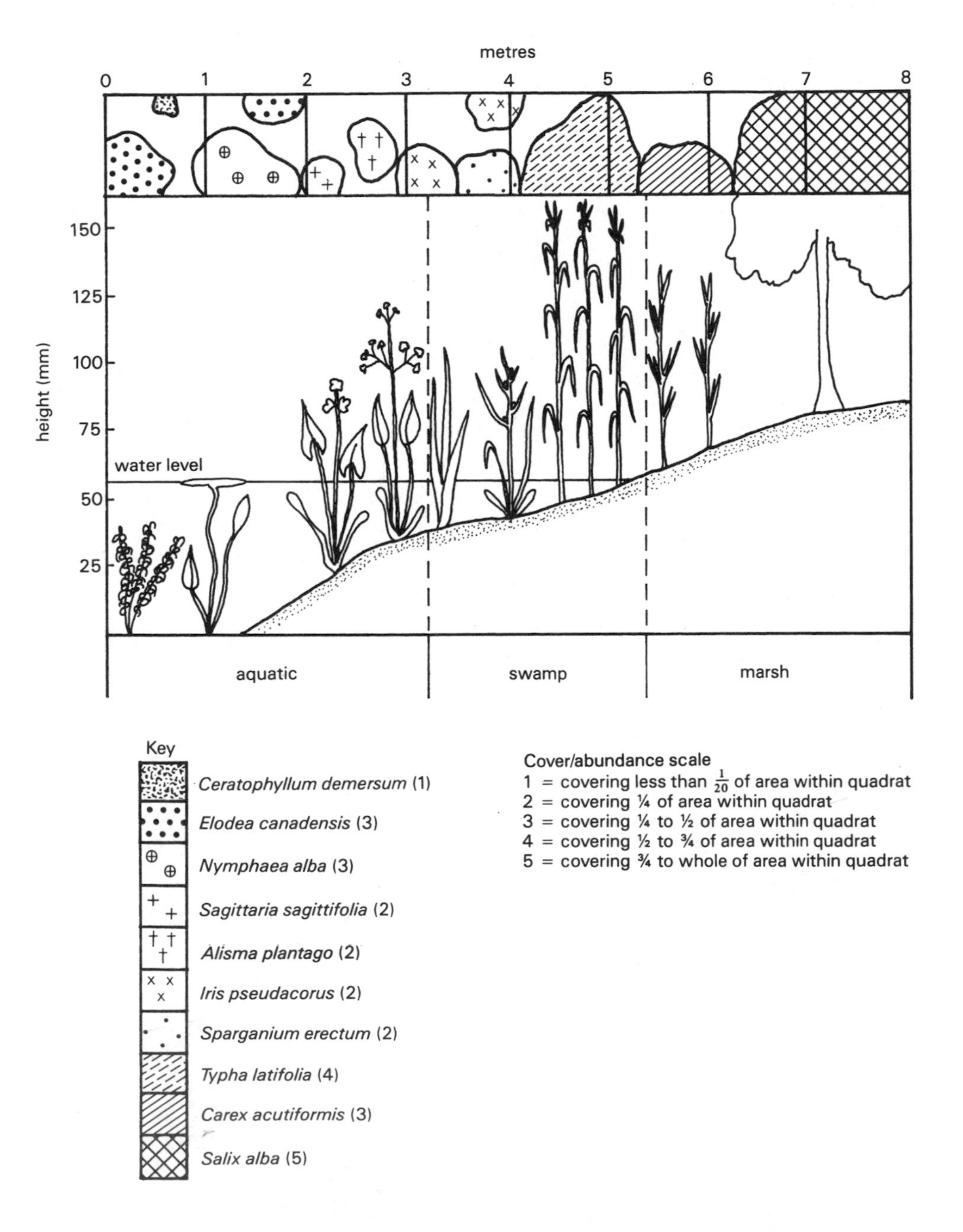

Fig. 4.5 Hypothetical belt transect and profile chart showing sequence of plant succession from water to marsh and dry land (hydrosere)

5 Plant communities of still water

The extent of a hydrosere (p. 27) is seldom clearly defined. Fringing plant communities frequently extend from dry land through an area of marsh and reed swamp to the open water. Typical dominant plants of swampy land are the bull rush or reedmace, *Typha* sp., and the common reed, *Phragmites communis*, which often reach a height of 2.4 metres or more, covering large areas with dense growth. When the reeds die they form a thick layer of rotting vegetation which builds up in height every year. Such dense material gradually encroaches on the open water and, if unchecked, will eventually form a marsh.

The many species of fringing and aquatic angiosperms can be grouped according to their manner of growth:

(i) Weeds rooted in the substrate but with leaves and flowers in the air. To this group belong the larger plants with rigid, upright stems (the bur-reeds, *Sparganium* spp.; water plantain, *Alisma plantago-aquatica*; and arrowhead, *Sagittaria sagittifolia*).
(ii) Weeds rooted in the substrate but with floating leaves (water lilies, *Nymphaea* spp.; floating pondweed, *Potamogeton natans*).
(iii) Weeds not rooted in the substrate and floating beneath the surface (water violet, *Hottonia palustris*; hornwort, *Ceratophyllum submersum*).
(iv) Weeds floating on the surface (all the species of duckweed, *Lemna* sp; frogbit, *Hydrocharis morsus-ranae*).

In addition to the above Angiosperms there are other groups of macrophytes: the stoneworts (Charales), liverworts (Hepaticae), a few species of water moss (Musci), the water fern (*Azolla*, Plate 5.1), quillworts (Isoetaceae), pillworts (Marsileaceae), and the horsetails (Equisetaceae).

Plate 5.1 Water fern, *Azolla filiculoides*, a true fern introduced to this country from the United States. Tends to spread rapidly under suitable conditions. Colour variable from pale green to bright red

Freshwater algae abound in both numbers and species and some have been mentioned in Chapter 3. Hosts of bacteria, algae and fungi grow upon the stems and leaves of the fringing vegetation and form the food of many species of grazing invertebrates (Fig. 5.1 (a)). Some are attached by stalks or mucilage pads, others are not directly attached but may be embedded in slime. As the community develops, becoming denser, filamentous algae grow and become draped from plant to plant. Within this sheltered underwater forest, flagellates can move about in their millions, undisturbed by currents.

Form and adaptability of plants

The high density of fresh water (700 times that of air at normal temperature and pressure) increases the buoyancy of aquatic plants so that rigidity is not so necessary as it is for land species, except in the aerial stems of emergent plants, such as the bull rushes and reedmace. These show considerable secondary thickening, whereas in the stems and leaves of many aquatic species, mechanical tissue is much reduced. The reduction in woody tissue is evident if a submerged plant such as water starwort, *Callitriche* sp., is lifted out of the water. In air its stems are totally unable to support themselves in an upright position.

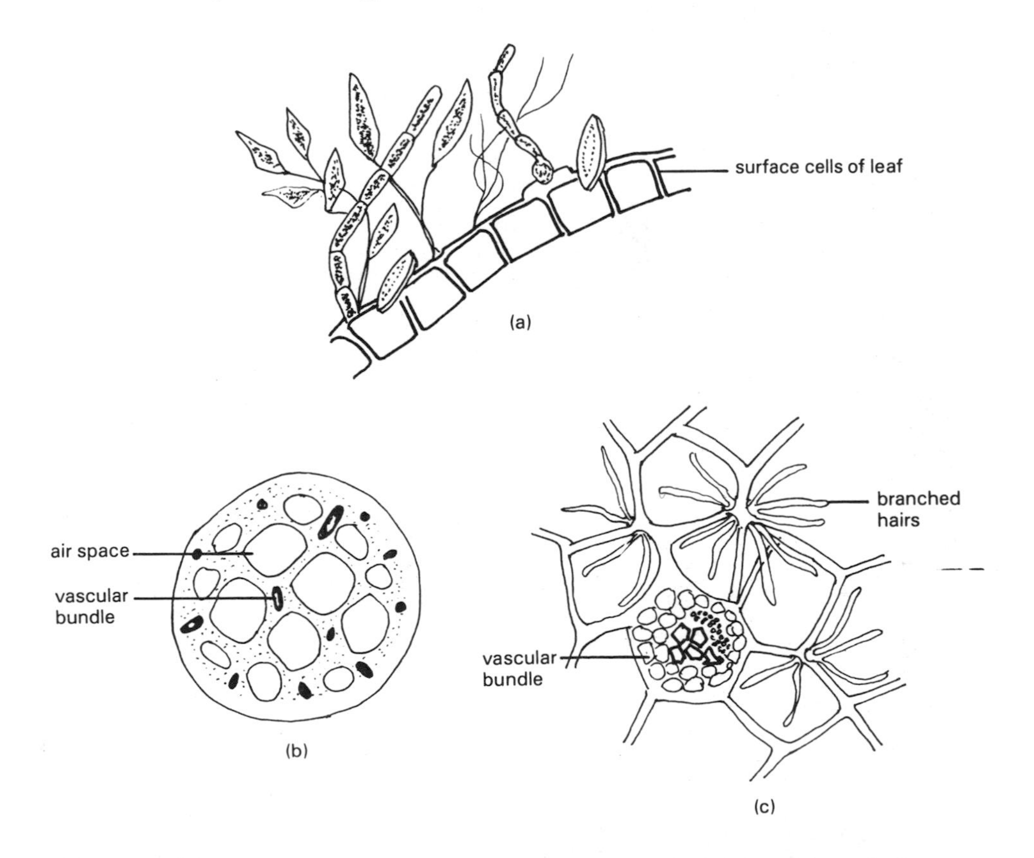

Fig. 5.1 (a) Epiphytic algae, fungi and bacteria growing on the surface of an underwater leaf. (b) Cross-section of the leaf petiole of white water lily, *Nymphaea alba*, showing the numerous air spaces and the scattered vascular bundles typical of many aquatic angiosperms. (c) Cross-section through leaf of water lily magnified to show the branched hairs within the cells which support the cavities in the mesophyll leaf tissues

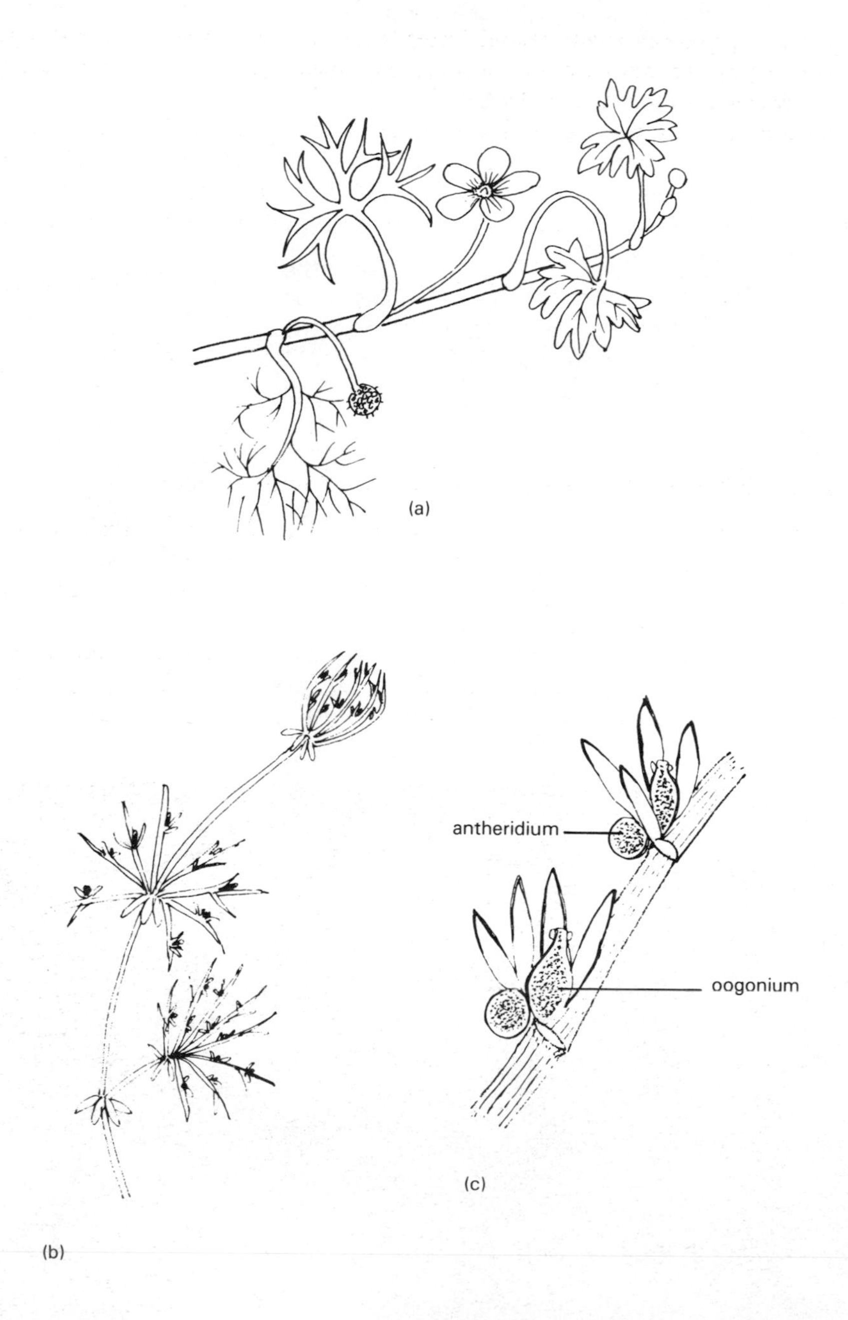

Fig. 5.2 (a) Water crowfoot, *Ranunculus aquatilis*, found in both still and slow-flowing water. The floating leaves help to support the flowers at the surface. Underwater leaves are finely divided. (b) Stonewort, *Chara* sp. Stoneworts are found in hard waters and often become encrusted with deposits of lime. (c) One of the side shoots of *Chara* enlarged to show the reproductive organs. The spherical male antheridium and pear-shaped oogonium are visible as blackish dots to the naked eye

Another important structural feature is the presence of numerous air spaces in both leaves and stems (Fig. 5.1 (b,c)). These not only give buoyancy but also act as storage areas for gases (p. 43).

Buoyancy in many species with floating leaves, such as *Potamogeton natans* and water lilies, is assisted by the waxy covering to the leaves which renders the upper surface unwettable thereby assisting them to float.

Leaf shapes in aquatic plants vary enormously, even within a single individual. Floating leaves are usually broad, sometimes forming a rosette as in water starwort. The underwater leaves of this plant are reduced to a thin strap shape. The water crowfoot, *Ranunculus aquatilis* (Fig. 5.2 (a)), also have leaves of two types: expanded floating leaves and underwater ones which are extremely finely divided.

As a plant grows the shape of its leaves may change. This is well illustrated by arrowhead (Fig. 5.3 (a)) in which the lowest leaves, developing first and subject to underwater currents, are narrow and grass-like. As growth proceeds towards the surface, ovate, floating leaves are formed, while the last to develop are arrow-shaped, emerging above the surface. (Plate 5.2).

Reproduction and seasonal adaptations

Aquatic angiosperms, like all flowering plants, must be able to achieve pollination. The floating leaves of *Potamogeton natans* support the flower spikes above the surface thus keeping the ripe pollen dry, a device used also by water lilies, water starwort and many others.

For those aquatic plants living below the surface, such as Canadian pondweed (*Elodea canadensis*), water violet (*Hottonia palustris*), and bladderwort (*Utricularia* sp.), devices for successful pollination are required. *Hottonia* (Fig. 5.3 (b)) grows a thick mat of finely

Plate 5.2 Arrowhead, *Sagittaria sagittifolia*, with a flowering spike growing out from the centre of the emergent leaves

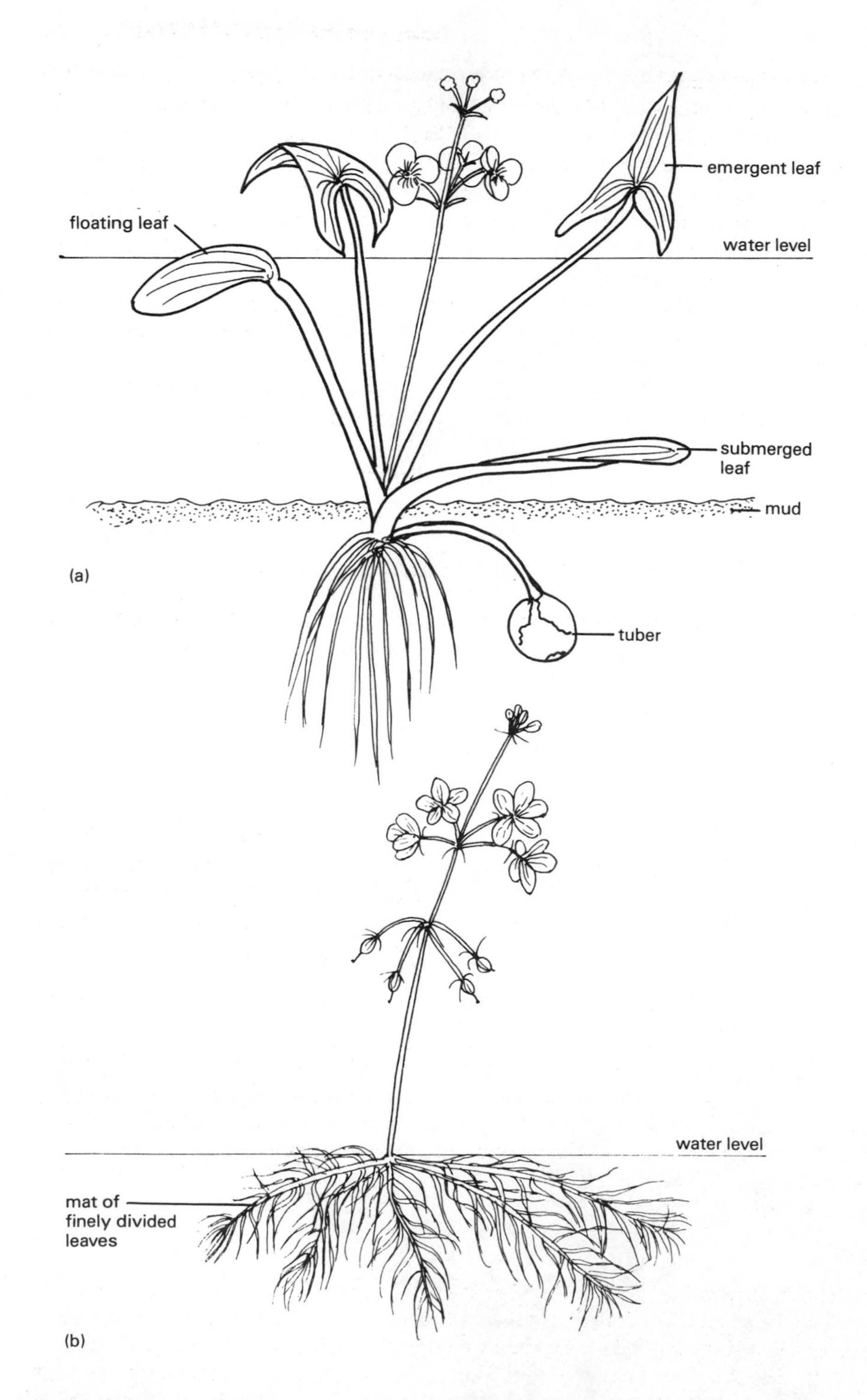

Fig. 5.3 (a) Arrowhead, *Sagittaria sagittifolia*, showing different leaf shapes. (b) Water violet, *Hottonia palustris*. The mat of finely divided leaves just below the surface supports the aerial flowering spike

divided leaves just beneath the surface while the flowering spike arises from the centre of the mat, which supports the stem, bearing whorls of pale mauve flowers above the surface. In a similar manner the fine underwater leaves of bladderwort support the flowering spike, buoyancy also being assisted by the bladders. Canadian pondweed rarely flowers but occasionally a very long thread-like flower stem arises from a leaf axil bearing a minute white flower at the surface (Fig. 5.4 (a)). Male flowers of this pondweed are also rare but when they are produced, the buds float upwards and burst to liberate pollen which blows along the surface to pollinate the female flowers.

Many aquatic angiosperms also reproduce vegetatively. Canadian pondweed produces specialized side shoots with closely packed leaves. These are called **turions** (Fig. 5.4 (a)). Eventually they break off from the parent plant and grow to form a new individual. This type of reproduction is common among aquatic angiosperms and turions are formed in summer and as over-wintering structures. Frogbit (*Hydrochan's morsus-ranae*) produces underwater turions in the form of tight buds (Fig. 5.4 (b)). In autumn these sink to the bottom and the rest of the plant dies completely. The buds persist on the muddy bottom and, in the spring, float to the surface and open out to form the first leaves. Growth is then rapid, more leaves being formed as well as long roots which hang in the water. Side shoots are also produced and soon a ditch can become closely covered.

Other methods of vegetative reproduction are tubers, like those of arrowhead, which persist overwinter in the mud to produce new plants in spring. The water soldier, *Stratiotes aloides* (Plate 5.3), produces side shoots at the end of which small plantlets develop. In a short time a stretch of water can become covered with these plants.

Physiological problems

Plants living in or near water, unlike their terrestrial relatives, are in little danger of suffering drought conditions when transpiration exceeds water intake. However, plants living in swamps and marshes can suffer serious water loss by transpiration from the great surface area of leaves projecting above ground. To counteract such loss they adopt various protective measures like the tough cuticle found in reeds and rushes and the sunken stomata of various grasses such as *Glyceria* sp.

Depth of water

The fringing plant life of lakes decreases with depth due to the rapid absorption of light rays by water. Vascular plants, with large air spaces within their tissues, suffer from changing hydrostatic pressure. At a depth of 10 m, the pressure of less than one atmosphere is sufficient to inhibit root growth and, significantly, vascular plants are rarely found growing below a depth of 10 m.

Light

Without doubt light is the most important factor in controlling the distribution of submerged aquatic plants. The macrophytes cannot penetrate to the depths of the euphotic zone at which phytoplankton is found. The bulky nature of these larger plants causes shading of the parts beneath them and the reduction of light intensity consequently reduces photosynthesis to a level where it equals respiration. This is called the **compensation point**, and lies at higher photosynthetic rates and light intensities than that for micro-algae. Stoneworts and mosses can withstand greater depths provided that the water is clear.

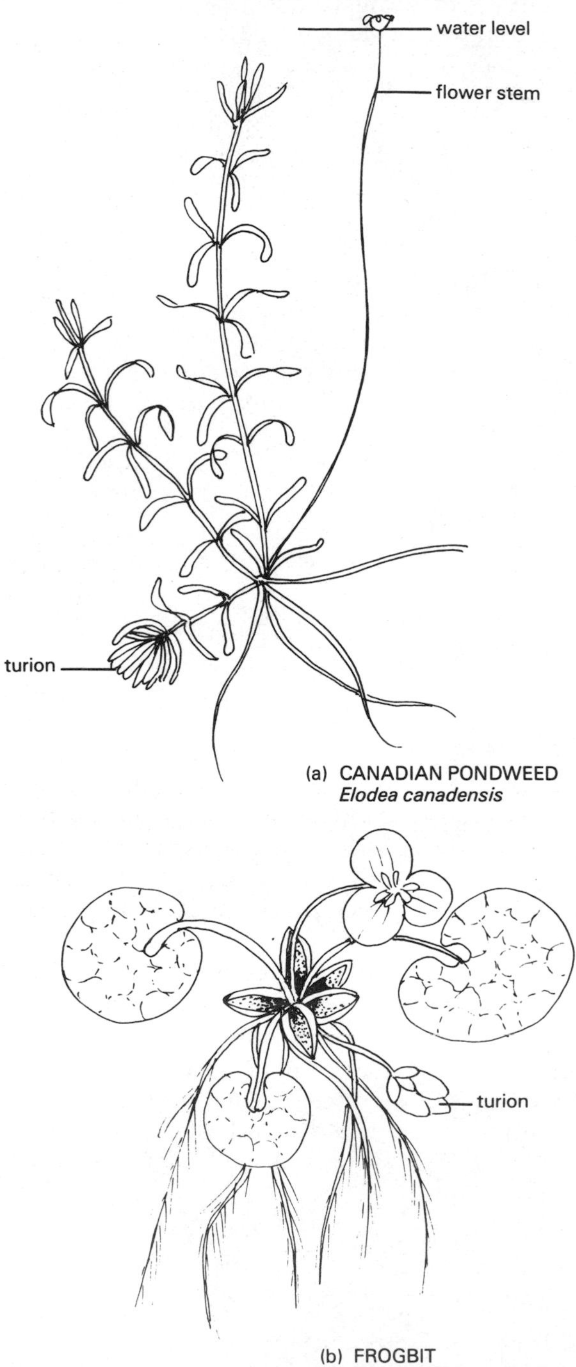

Fig. 5.4 (a) Canadian pondweed, *Elodea canadensis*, bearing a long, thread-like flower stem to bring the female flower to the surface for pollination. (b) Frogbit, *Hydrocharis morsus-ranae*. A well-developed floating plant bearing a flower and turion

Nutrients

Fringing plants that are partly aerial and partly submerged can make use of atmospheric carbon dioxide as well as the copious supply of dissolved nutrients in the water circulating round their root systems. These nutrients are constantly being replenished by the deposition of rich sediments. No wonder that a garden pond quickly becomes colonized with a variety of water plants to the point where drastic thinning out becomes necessary. Vascular plants are at an advantage compared with non-vascular aquatics since their root systems can make use of water contained in the pore system of the sediments. This pore water contains something like ten times more nutrients than that of the overlying waters. Algae living on the surface of the sediments also make use of nutrients diffusing upwards from the pore water. This results in a rich bloom of algal growth.

Of all the nutrient salts present in fresh water, phosphorus is probably the most important in governing the productivity of fringing plants. The great solubility of phosphorus compounds in oxygen-free sediments increases their availability.

Nitrates are used by aquatic plants for the formation of proteins. During the period of rapid growth in the spring, much of the nitrate content of the water is removed, to be restored in the autumn by bacterial breakdown of decaying plant and animal material. The bladderworts, floating submerged beneath the surface, have no true roots and live in peaty pools poor in mineral salts. However, they make good this deficiency in another way. Small bladders are borne on the hair-like leaves. Each bladder has a circular aperture at the end surrounded by a funnel of bristles and closed by a trap-door which opens inwards (Fig. 5.5). Glands on the inside wall of the bladder extract cell sap. This causes a tension when the trap-door is closed and the bladder contracts. Protozoans, rotifers and small crustaceans, possibly attracted by mucilage, touch the sensitive bristles surrounding the opening. This causes the release of the trap door and the bladder fills rapidly with water, drawing in the microscopic animals. After dying within the bladder, their decomposed remains are absorbed by the plant, in this way compensating for the lack of minerals, and particularly of nitrates.

Calcium carbonate also diminishes markedly during the season of greatest plant activity. The plants extract carbon dioxide and the soluble bicarbonate from the water leaving the insoluble carbonate as a precipitate. In waters rich in calcium this precipitate can be noticed on the leaves of water plants.

For most of the year, water soldier plants float just beneath the surface, their long roots hanging down in the water to keep them in a level position. If removed to another place in

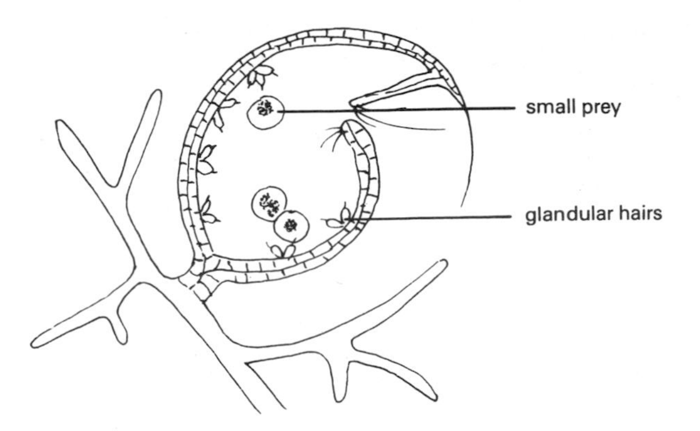

Fig. 5.5 Bladderwort, *Utricularia* sp. Cross-section through one of the bladders much enlarged to show the internal structure

the pond they will heel over until they have regained their level position. In early summer, young leaves are produced which increase the air-space/tissue-volume ratio and hence their buoyancy, so that the plants rise to bring the new leaves above the surface. White flowers appear at this time, arising from the centre of the plants (Plate 5.3), but in Britain they are all females and no seed is set. As we have already seen (p. 39) the plants reproduce vegetatively by growing plantlets on the end of stolons. In winter the plants sink once more beneath the surface.

Many plants can accumulate substances which they absorb from the water. Such a process takes place in the stonewort, *Chara* (Fig. 5.2 (b,c)). If a stonewort plant is placed in water containing 26 mg l^{-1} calcium ions, the cell sap will be found to contain as much as 380 mg l^{-1} of the ions. The ability of the stoneworts to accumulate salts of calcium accounts for their name and for the fact that their whorls of leaves are often covered with a heavy deposit of calcium carbonate.

The fluctuation in the concentration of salts is greatly influenced by plant activity which in the end rebounds on the plants themselves, for depletion of the salts checks multiplication until decay once more restores the balance. The seasonal activity of the plants and variation in the concentration of salts is a reciprocal process. The salts reach their highest level at a time of the year when growth is at its lowest and the level decreases with increased uptake by the plants when renewed growth begins.

Carbon dioxide

This gas, necessary for photosynthesis, can be obtained direct from the atmosphere, in solution in the water or from soluble bicarbonates. Aquatic plants also have an advantage

Plate 5.3 Water soldier, *Stratiotes aloides*. Note the white flowers produced in high summer when the plants are at the surface

over terrestrial species since the highly organic sediments, by their rapid decomposition, produce a large amount of carbon dioxide.

The part played by the air spaces present in aquatic species is important since they act as reservoirs of carbon dioxide, the result of diffusion from the sediments. The concentration of this stored carbon dioxide is many times higher than that in air. In addition, the air spaces also store the surplus carbon dioxide produced by respiration at night, assuring a plentiful supply with the onset of photosynthesis in the first hours of daylight.

Productivity of aquatic plant communities

The extent of fringing communities will depend upon the area of shallow water round a lake which is suitable for colonization. Such a community can be extremely productive, indeed among the most productive of all vegetation types. Growth of terrestrial plants is usually limited by lack of water and nutrients, neither of which is denied to aquatic macrophytes. These organisms are in the happy position of receiving all they require from the catchment area, which also brings in fresh silt. Algae, fungi and bacteria colonizing the macrophytes also contribute in no small measure to the total productivity of the community.

Productivity depends on the effects of the physiological factors (Chapter 2) and upon the physical factors prevailing. Measurement of the productivity of these communities is fraught with many difficulties and the degree of accuracy will depend on the method employed for the measurement and collection of the plants. Some estimates are based on the dry weights of samples within the community but this can be subject to error since the amount of ash in plant tissue varies enormously with different individuals. More accurate results are obtained by recording the organic content, that is the ash-free dry weight.

In harvesting, the underground parts of the plants are often not included and these can account for up to 50 per cent of the total plant production. Again, records of standing crop are often made at the period of greatest growth, which does not take into account leaves or stems which have grown and died before or after records were made, thereby leading to a false estimate of total annual production. Although doubts have been cast as to the accuracy of results obtained by whatever method used, a general picture emerges of aquatic plant productivity which is probably approximately correct. Figure 5.6 summarizes the results of many measurements made by Hall and Moll [5.1] and by Westlake [5.2]. From this it is evident that emergent aquatic plants form some of the most productive plant communities of all vegetation types.

Plants and the freshwater community

As primary producers, aquatic plants, by their photosynthetic activities, are the main source of oxygen. Fluctuations occur in the oxygen content of any body of water. Apart from the effects of physical factors mentioned in Chapter 2, oxygen is only manufactured by plants during daylight and the amount produced is dependent upon the quantity of light energy which will vary between dawn, midday and dusk and also with the depth at which the underwater leaves are growing. The released oxygen quickly dissolves in the surrounding water so that under optimum conditions, this may result in areas of water in the proximity of aquatic plants becoming saturated with oxygen or even supersaturated. During the hours of darkness there is a fall in the amount of dissolved oxygen. Besides these diurnal fluctuations, there are seasonal changes as in winter when the plants die and oxygen production falls. To a certain extent this fall will be offset by the lower temperature of the water in winter when the dissolved oxygen level can rise.

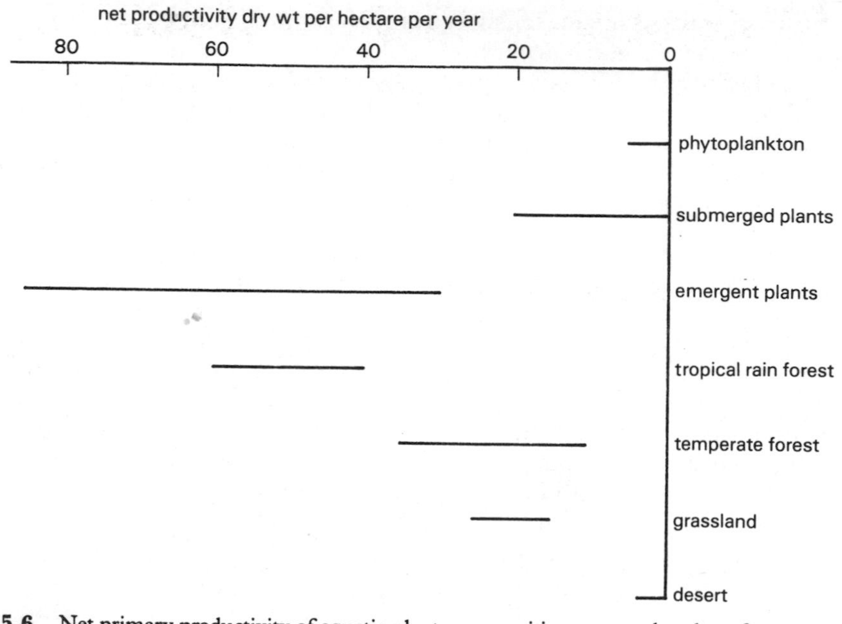

Fig. 5.6 Net primary productivity of aquatic plant communities compared to that of phytoplankton and terrestrial vegetation (summarized from Westlake, 1963)

As well as being the essential providers of oxygen, plants also offer a means of attachment for those organisms which by their buoyancy would otherwise float to the surface. They also provide shelter for innumerable small animals. Many animals, such as the water stick insect, water scorpion, dragonflies and mayflies use the underwater leaves of plants on which to lay their eggs (Fig. 11.7). The gill-breathing nymphs of dragonflies and mayflies climb up the aerial stems of plants before emerging as air-breathing adults.

Finally, after death, the decaying plant bodies serve as a rich source of food for numerous species of detritivores, chief among which are fly larvae, worms, and crustaceans such as the water louse, *Asellus* sp.

Fieldwork

1. In the belt transect constructed in the fieldwork for the last chapter, examine the rooting systems of the plants recorded within the transect to find out how they over-winter. Which plants die down altogether?
2. Examine some of the bladders produced by bladderwort, *Utricularia* sp., under the low power of a binocular microscope and identify any animals you can see.
3. Using a Secchi disc (Chapter 2, p. 11 and Fig. 2.6), determine the depth of visibility in different regions of a pond or lake. How do these measurements relate to the growth of water plants at the sites chosen for measurement?

References

5.1 Hall, C.A.S. and Moll, R. (1975) Methods of assessing aquatic primary productivity, in *Primary Productivity of the Biosphere,* (eds) Lieth, H. and Whittaker, R.H., pp. 19–54, Springer–Verlage

5.2 Westlake, D.F. (1963) 'Comparison of plant productivity', *Biol. Rev.*, **38**, 385–425

6 Animal communities of still water

The variety of habitats to be found in still waters is only now being realized as modern methods of lake survey and more sophisticated ecological studies become possible. Nevertheless, the way in which habitats are defined and classified remains difficult, especially as they are so variable, ranging from temporary puddles to small ponds, and from bogs to large lakes which can be both shallow and deep. The whole subject is further complicated by the fact that each type of habitat comprises a number of microhabitats.

Open waters, such as lakes and large ponds, comprise three main regions which are illustrated in Fig. 6.1. The **pelagic region** is the area of free water away from the influence of the shore and the bottom substrates. The **benthic region** is the bottom area extending from the shore to deeper parts. The shallower zones include the surface of the sediments existing among the beds of aquatic macrophytes and are therefore within the euphotic zone (area of water receiving light). Here the sediments provide habitats for a large number of species. The **profundal region** is that below the depth of the illuminated zone and the inhabitants must rely for their food upon organic matter produced in the euphotic zone. They are therefore mostly decomposers and as such are active members of the detritus chain. Within these three regions there are communities of animals specially adapted for life in these habitats.

Associated with the surface of the water are animals known collectively as the **epineuston**, living on the upper surface. Those living on the lower surface film are **hyponeustic** animals, known collectively as the **hyponeuston.** The **nekton** (from the Greek meaning moving) includes all those animals such as fish and many invertebrates which move actively between the pelagic and benthic regions. The mobility of the nekton enables them to remain in zones of high productivity and also where they are largely immune to currents which might otherwise sweep them away. Their mobility also means that they can select favourable habitats in which to pass different periods of their lives by migrating from one area to another.

The **benthos** consists of organisms associated with the bottom sediments found in the littoral and profundal regions. The large number of microhabitats present in these areas

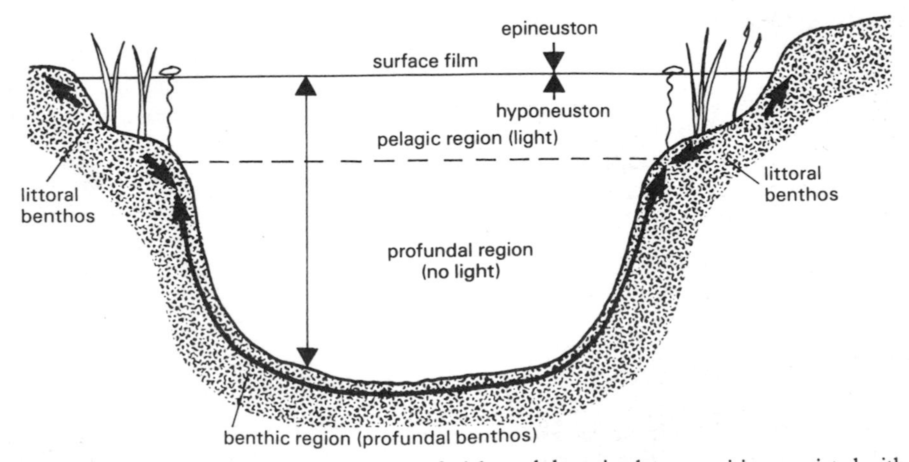

Fig. 6.1 Diagram to show the main regions of a lake and the animal communities associated with these regions

afford opportunities for a variety of bottom-living animals, many of which are deposit- and filter-feeding creatures.

Animals associated with the surface film

There are several species of water bug, belonging to the order Hemiptera, which are epineustic. They spend their entire lives on the surface performing skating movements and feeding on small insects which fall on the surface from surrounding vegetation. The pond skater, *Gerris* sp. (Fig. 11.1 (a)), is probably the commonest but there are others such as the water cricket and the water measurer.

These surface dwelling, predatory hemipterans form a close community and make the link between the world above water and that below the surface. They themselves are preyed upon by the nekton, notably fish, and even by other hemipterans such as the water boatman.

One of the water beetles, the whirligig beetle, *Gyrinus natator*, also lives on the surface (Fig. 11.1 (b)) but it is not totally confined to the surface and is able to dive below when danger threatens and even to fly from one pond to another.

Among hyponeustic forms can be counted the microscopic rhizopod, *Arcella* (Fig. 3.2), which secretes a shell of chitin. *Arcella*, although usually to be found on the surface of rich mud, can also dwell hanging beneath the surface film of pond water along which it glides feeding on particles of matter adhering to it.

Fish as part of the nekton

The nekton contains a wide range of animals, including fish, which play a vital role in this community. Although a great deal has been written about the natural history of freshwater fish, many details of their habits and even of their life history is still unknown.

Through predation fish can affect plants and animals far removed from them in the food chain. The problems arising from man and his fisheries are dealt with in Chapter 13. Here we are concerned with freshwater fish as part of the nekton, along with invertebrate members of this community.

Many of our freshwater fish spend part of their lives in lakes and ponds, and part in the flowing waters of rivers and streams. A few, like the salmon, spawn in fresh water, the young fish spending the first part of their lives there, while part of the life cycle is spent in the sea, to which they migrate via rivers and estuaries. Such fish are said to be **anadromous**. Eels, on the other hand, are often **catadromous**, for they spawn in the sea and migrate up rivers to spend their larval and adult lives in fresh water. In both, the ability to migrate offers opportunities to exploit different habitats for food at different periods of their life cycles (Chapter 10).

Tolerance to environmental factors and the effects of predation and competition determine the distribution of fish in fresh waters. From Table 6.1 it is clear that tench and carp thrive at higher temperatures than do salmon or trout. But although such environmental factors obviously influence choice of habitat, as between cold and warmer waters, the data do not explain why one species may predominate in a particular lake.

Invertebrate inhabitants of weed beds

The floating and submerged vegetation of many ponds in summer can be so extensive as to form a complete covering to the water, thereby reducing the light penetration. The underwater stems and leaves of aquatic macrophytes can be heavily colonized by epiphytic

Table 6.1 Temperature and oxygen requirements for growth and spawning of some species of British freshwater fishes (after Varley (1967). *British freshwater fishes.* Fishing News (Books) Ltd., London). Data for salmon and minnow is incomplete.

Fish	*Optimal growth temp.* (°C)	*Upper lethal temp.* (°C)	*Spawning temp.* (°C)	*Usual dissolved* O_2 *requirement* (mg l^{-1})
Salmon		32–34	2–6(–10)	
Trout	7–17	26·5	(→10)	10–16
Pike	14–23	29	10+	5–6
Minnow				10–16
Perch	14–23	32·8	10+	5–6
Roach	14–23	33·5	10+	5–6
Tench	20–28	35·2	15+	0·7+
Carp	20–28	37·0	15+	0·7+

organisms (**periphyton**) such as algae, bacteria and protozoa (Fig. 5.1 (a)). They also acquire a covering of detritus. This rich substrate forms the feeding ground of many grazing animals which cling to the weeds, such as snails, chironomid larvae, the nymphs of the mayfly, *Chloeon dipterum* and many species of caddis larvae. In fact it is true to say that most of the weed-bed herbivores feed by grazing the periphyton rather than eating the plant tissues themselves. A great variety of small crustaceans, such as cladocerans, ostracods and copepods (Figs. 6.5 (a) and 3.3), feed by filtering detached fine material as they swim around or cling to the weeds. The water louse *Asellus* (Fig. 6.2 (c)), feeds on organic detrital matter among the plants and on the mud. The misnamed river sponge, *Spongilla fluviatilis*, (Fig. 6.2(a)) which is really a still-water species, encrusts the stems of the macrophytes. Sponges feed by taking in a current of water through small pores, from which particles of food are filtered out. In turn, freshwater sponges offer shelter to a number of small animals, notably the larvae of the parasitic spongilla fly, *Sisyra* sp. (Fig. 6.2 (b)), which suck the juices inside the cavities of the sponge.

Amongst the carnivores of the weed beds there are the active hunters and what may be termed the lurkers. The latter are mostly totally aquatic species, being unable to use atmospheric air, and depend upon a supply of dissolved oxygen from the water. This is either absorbed directly through the body integument or gills of various kinds are used for its extraction (Chapter 11).

The lurkers rely on remaining still and often concealed not only from their own prey but also from other predators. Coelenterates such as *Hydra* sp., and the less frequent hydroid, *Cordylophora lacustris* (Fig. 6.2 (d,e)), remain attached to a plant stem or leaf and extend their tentacles, equipped with sting cells, to capture water fleas and other small animals.

Several species of freshwater leech inhabit the weed beds, attaching themselves by means of their powerful posterior suckers and stretching out their extensible bodies so that they sway to and fro in the water. Some species attack passing invertebrates, others are blood suckers parasitizing fish and waterfowl. *Pisciola geometra* (Fig 6.3(a)), the fish leech, is often plentiful among underwater vegetation, its range being limited to well-oxygenated water. This leech has a large anterior sucker by means of which it attaches itself to passing fish, piercing the skin of the gills or at the base of the fins with its sharp circlet of teeth. *P. geometra*, although fairly widely distributed, is not common except in waters with a good fish population and can be a serious pest not only by causing loss of blood to the host but also because the skin abrasions, left by the teeth, offer easy entry to bacterial and fungal diseases.

Fig. 6.2 Some invertebrate inhabitants of weed beds. (a) River sponge, *Spongilla fluviatilis,* (Porifera) encrusting plant stem. (b) Larva of sponge fly, *Sisyra* sp. (Neuroptera) which lives within the cavities of the sponge. (c) Water louse, *Asellus* sp. (Isopoda) feeds on organic matter. (d) and (e) *Cordylophora lacustre* (Coelenterata) forms branched colonies attached to underwater objects. Bears numerous hydra-like polyps. (f) *Corophium lacustre* (Amphipoda) lives in mud tubes among colonies of *C. lacustre* or on submerged plants. Feeds on organic matter. Body not laterally compressed. Note stout 2nd antennae

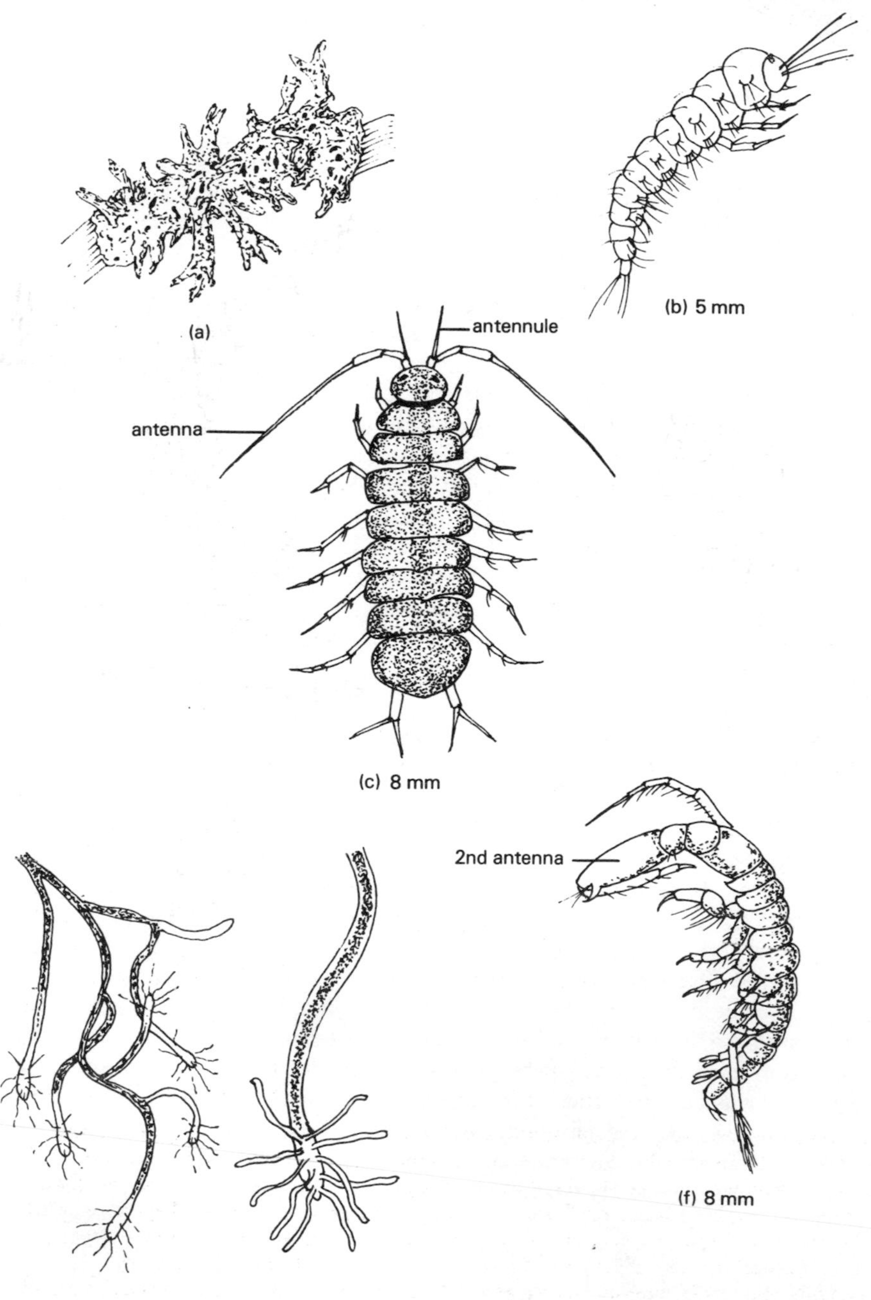

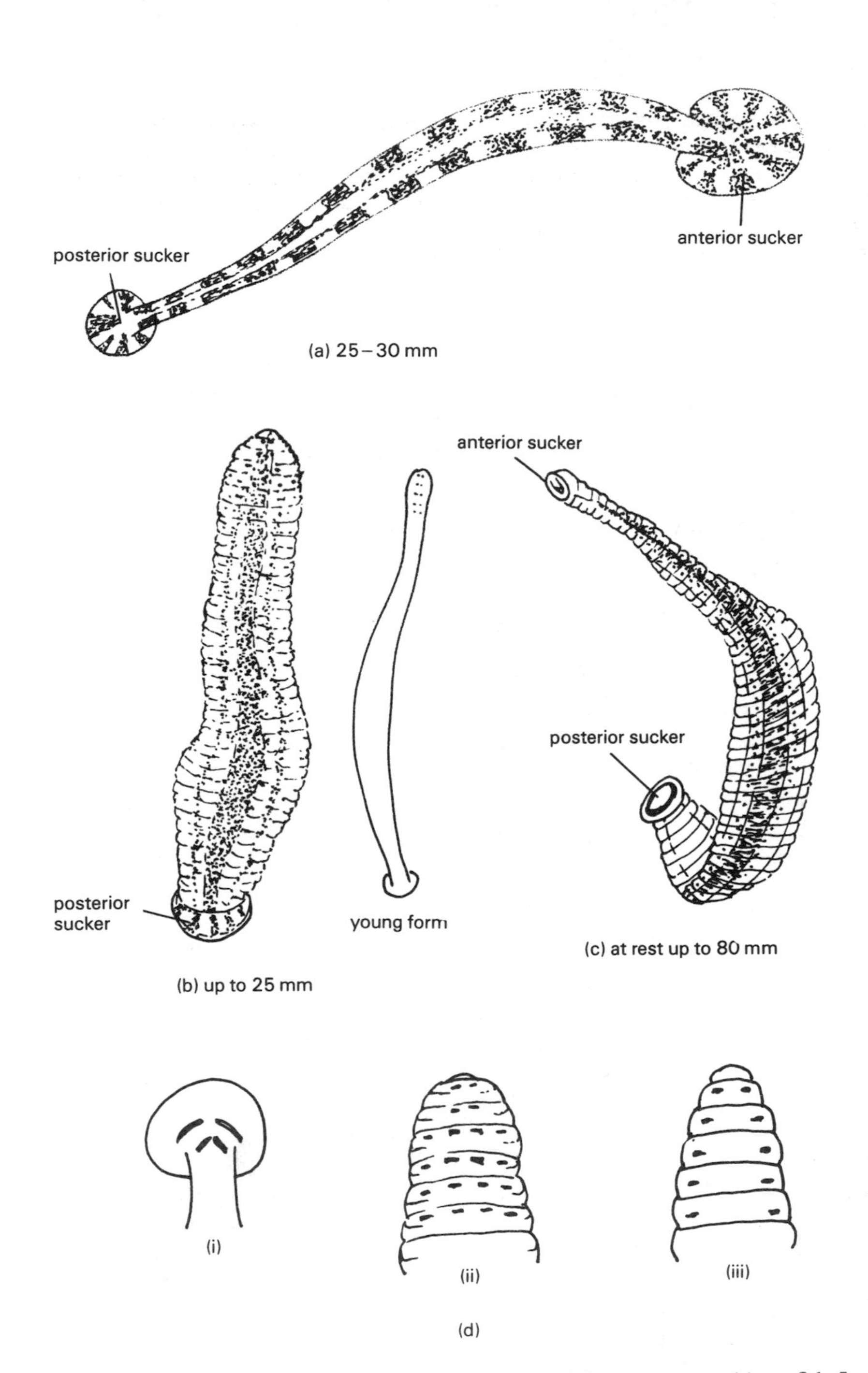

Fig. 6.3 Some leeches encountered in weed beds. (a) *Pisciola geometra*, parasitic on fish. Large anterior sucker. Cylindrical body which never contracts very much. Numerous where fish are plentiful. (b) *Theromyzon tessulatum*, parasitic in its young stages in the nasal cavities of waterfowl. Anterior sucker not visible from above. Posterior sucker conspicuous. Very active. Shape varies with degree of contraction. (c) Medicinal leech, *Hirudo medicinalis.* The largest British species. Parasitic on mammals. Colourful with red, yellow and black rows of dots. (d) Heads (i) *P. geometra*, (ii) *T. tessulatum*, and (iii) *H. medicinalis*, enlarged to show arrangement of eyes. These can best be seen when the head end extends

The medicinal leech, *Hirudo medicinalis* (Fig. 6.3 (c)), is the largest and most colourful of our British leeches. Like others of its kind, it inhabits well-weeded stretches, lurking in wait for vertebrate prey. Due to over-collecting for medicinal purposes in the nineteenth century, it became very scarce, many thousands of leeches being exported for use on the Continent where they were used to reduce contusions and for other medicinal purposes after the fashion for 'leeching' in this country had ceased. Now, however, there appears to be a revival in the use of leeches for certain post-operative conditions and leech-farming has begun once more in order to satisfy demands. The New Forest is one of the last strongholds of *H. medicinalis* in the wild where it parasitizes cattle and ponies wading in the water. The adults are so large that, when fully extended, they resemble a small snake as they swim along near the surface with an undulating motion.

One other species of leech must be mentioned here. This is *Theromyzon tessulatum* (Fig. 6.3 (b)) which, in its young stages, parasitizes waterfowl. Once attached to the beak of a water bird, the leech enters the nasal cavity and inserts its proboscis through the wall of the cavity. Several leeches feeding in this way can be fatal. When sated with its host's blood, the leech drops off and grows rapidly. Apparently three meals interspersed with growth periods are required before maturity is reached but the period between meals can vary from a few days to nine months. Long periods of enforced fasting are typical of leeches that suck the blood of vertebrates.

Among hunters of weed beds which move about actively searching for prey are many freshwater beetles, probably the most voracious of which is the great diving beetle, *Dytiscus marginalis* (Fig. 6.4). Both adults and larvae prey upon nearly everything around them including young fish and the tadpoles of newts, frogs, and toads.

Weed beds not only offer food and shelter for these and many other invertebrates but also protection for their eggs and newly hatched young. Many species use the leaves and stems of weeds on which to deposit eggs (Chapter 11) using various means of attachment.

Besides the lurkers and hunters, only some of which are mentioned here, there are the opportunists which move around among the weeds hopefully seeking a meal or a ride. It is difficult to dip a net amongst thick weed in summer without bringing up a number of bright red mites or their blacker relatives (Fig. 6.5 (c)). They have four pairs of legs and belong to the small order Hydracarina. The body of the adult mite appears to be undivided but there is a small false head, or capitulum, containing a pointed beak with which the juices of prey are sucked into the body. The larvae, on the other hand, have only three pairs of legs and a very distinct capitulum. In most species the larvae are parasitic and attach themselves to their host, often a water scorpion or beetle, burying the capitulum in its body (Fig. 6.5 (d)). There they remain, and even pupate, continually sucking the host's body juices.

The caddis larva, *Trianodes bicolor* (Fig. 6.5 (b)), is often found swimming actively among the weeds. A less frequent adventurer is the fish louse, *Argulus foliaceus*, always found in waters where there are fish. A feeble swimmer, this crustacean is beautifully adapted by its flattened body and two large ventral suckers as a parasite, attaching itself to a passing host. Once attached it penetrates the flesh of the host with a sharp beak and extracts body fluids (Fig. 6.5 (e)).

Invertebrates associated with the bottom sediments

As we have seen on p. 45, the invertebrates of the bottom sediments belong to two categories: those living within the euphotic zone in conditions of rich nutrient production (which are typical of the areas round the roots of aquatic macrophytes) and the inhabitants of the deep (profundal) sediments, which exist in darkness and in cold and often anoxic conditions.

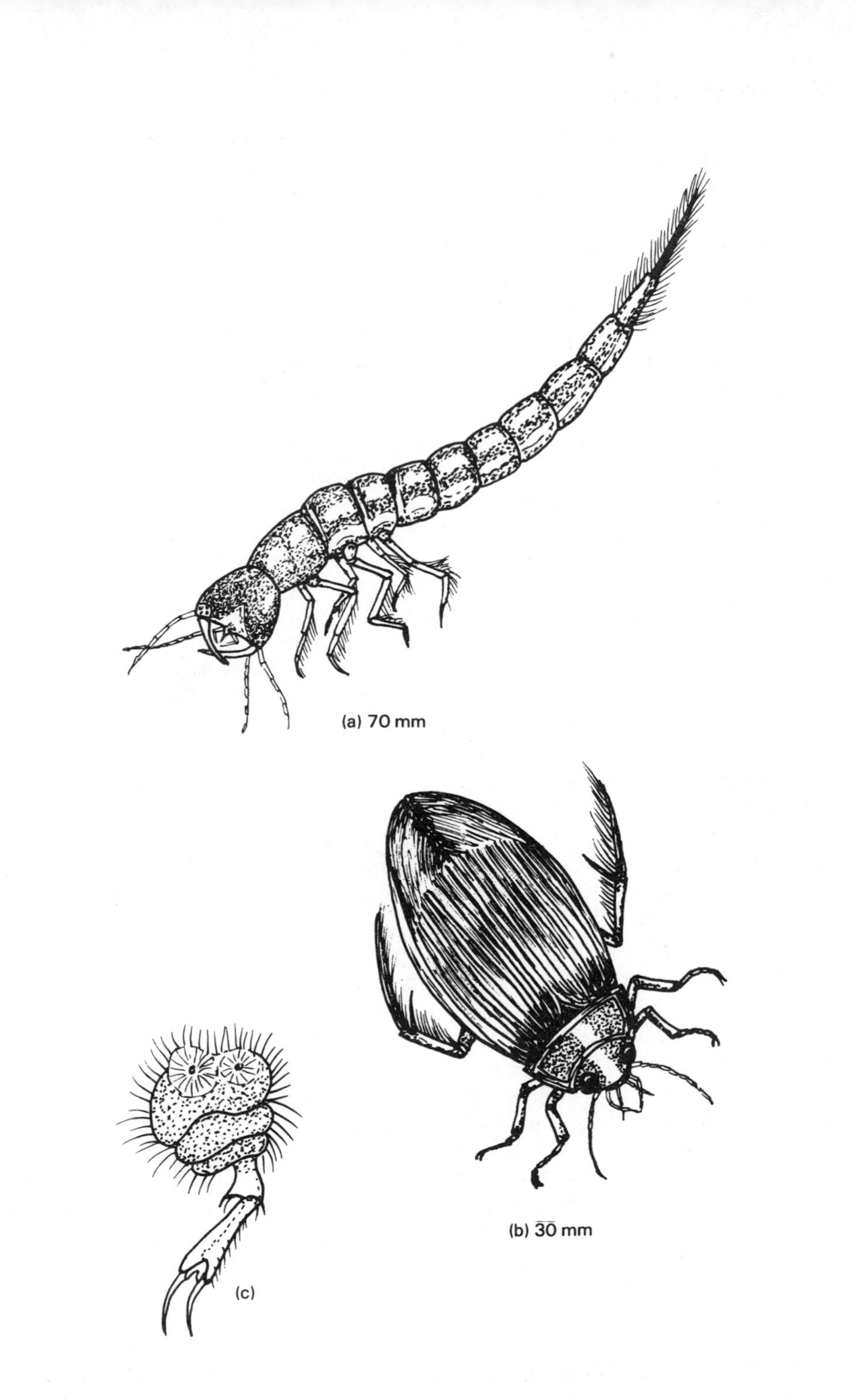

Fig. 6.4 A predator of the weed beds, the great diving beetle, *Dytiscus marginalis.* (a) Mature larva ready to pupate. (b) Female beetle. Note ridged elytra, which are smooth in the male, and the absence of discs on the forelegs. (c) Foreleg of male, ventral view. The disc is formed of three enlarged tarsal joints and helps to grip the female when mating

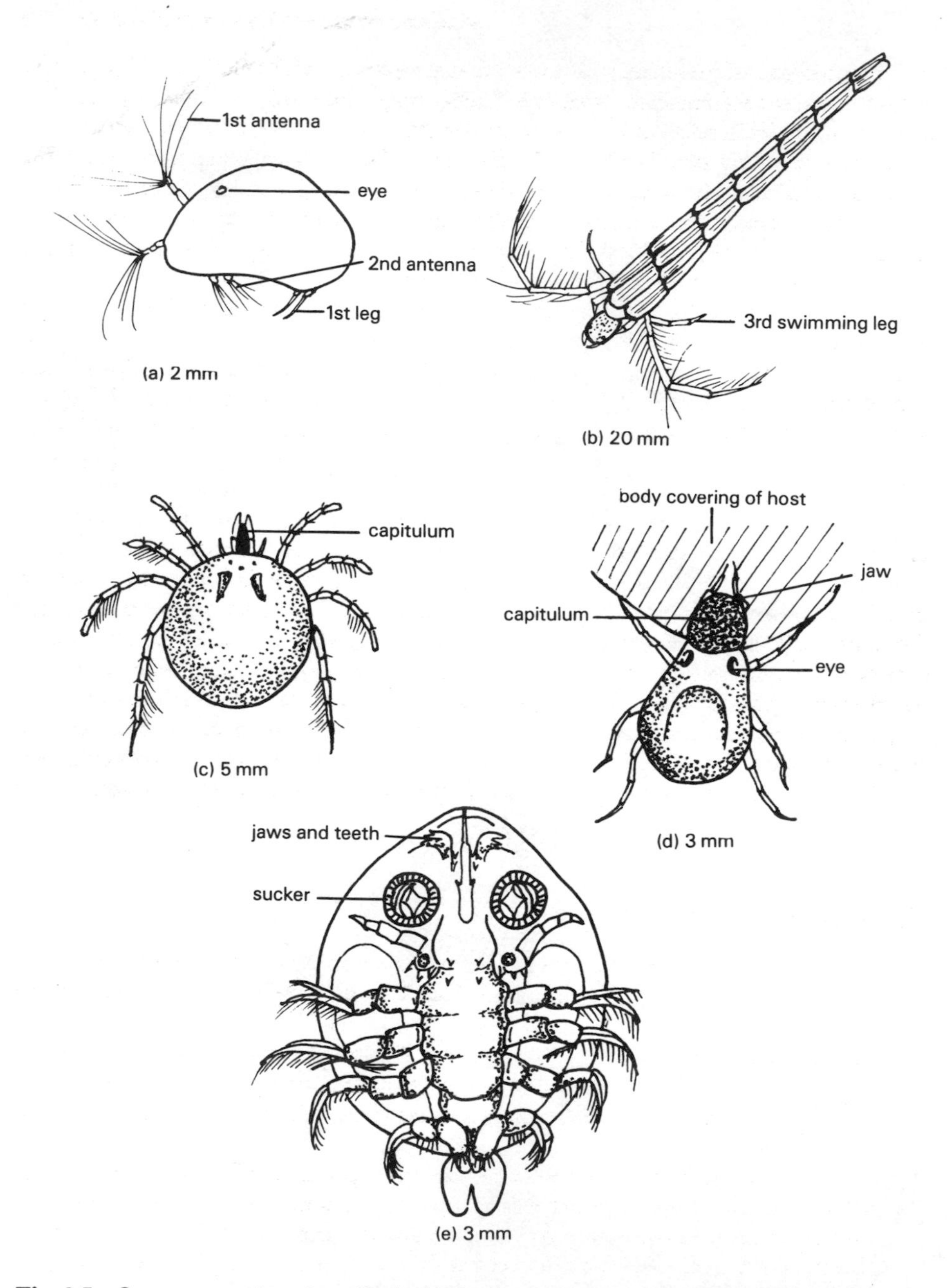

Fig. 6.5 Some opportunists of weed beds. (a) *Cypris* sp., an ostracod which swims among the weeds by moving its long antennae backwards and forwards alternately. The body is enclosed in a bivalve shell which obscures the internal structures. (b) The larva of *Trianodes bicolor*, a vegetarian caddis which swims actively by means of its long, hairy 3rd pair of legs. The case is constructed from pieces of reed or leaf arranged spirally. (c) *Hydrarachna* sp., one of the largest water mites, free-living and carnivorous as adults. (d) Hexapod larva of *Hydrarachna*, parasitic on a number of fresh-water insects; its strong jaws pierce the skin of the host and the false head or capitulum is embedded in the host's tissues. (e) Ventral view of fish louse, *Argulus foliaceus*. The extreme flattening of the transparent body offers little resistance to the water, thereby assisting the crustacean with its feeble swimming powers and, when attached to a host, preventing it from being brushed off

For the animals of the littoral sediments, the macrophytes and the mud around their roots provide a habitat for numerous species including oligochaete worms, flatworms, leeches, crustaceans, nymphs of certain species of mayfly and caddis, larvae of two-winged flies and a few bivalve and gastropod molluscs. By no means all of these animals spend their entire lives associated with the bottom mud at the edges of lakes and in ponds, for most of them move about to explore the weedy regions above and show no particular adaptations for an existence in or on the mud. There are, however, a few which are typical of this region and which possess special respiratory or feeding mechanisms (p. 124, Figs. 11.5 and 11.17).

The fauna of the profundal zone are largely decomposers, providing food for higher organisms such as fish, but they also liberate nutrients which affect the chemistry of the waters above. They exist in a less complex habitat than their counterparts of the littoral zone and, although the diversity of species is greatly reduced, their food supply, provided by the activities of the sedimentary bacteria, is virtually unlimited, hence they are present in large numbers.

Study of the animals living in these profundal regions is only possible by using special methods and equipment [6.1]. However, to complete the picture so far as we are concerned, mention of the important part played by the animals living in these sediments must be made. Operating often in anoxic conditions and virtually in complete darkness, they dwell in muddy sediments which continually receive additional fine deposits. Such a habitat demands a high degree of adaptability. Although the substratum offers a bountiful supply of food, they must be able to obtain and metabolize what they get. Thus they have developed specialist respiratory and feeding mechanisms and also often complicated life histories which may involve migration at different stages. A description of some of the adaptations adopted by these profundal organisms is given in Chapter 11. As in other ecosystems, the carnivores of this region are few compared to the large numbers of detritivores upon which they feed.

Fieldwork

1 In a pond you have selected for investigation, look more closely at the insects living on the surface. Make a list of the species you find. Which of them are able to dive beneath the surface?
2 Many aquatic insects possess water-repellent (hydrofuge) hairs. Observe (a) surface dwelling insects and (b) those insects normally living beneath the surface but which breathe atmospheric air. Of what particular use are hydrofuge hairs to each of these groups of insects?
3 Flatworms are attracted by raw meat. Compare the populations of flatworms in different parts of a pond such as on a muddy bottom, gravelly substratum, and beneath stones, by dropping small pieces of meat of the same size tied to a string and weighted with a stone. Haul the meat up after 24 hours and count the flatworms present. This will only give comparative results. Devise ways of obtaining more accurate results.

References

6.1 Schwoerbel, J. (1970) 'Methods of hydrobiology', *Freshwater Biology*, Pergamon Press
6.2 Vailey, M. E. (1967) *British Freshwater Fishes*, Fishing News (Book) Ltd

7 Flowing water of streams and rivers

The characteristics of fresh water — physical, chemical and biotic, described in Chapter 2, apply in lotic systems also. However, rivers and streams also possess characters of their own. This chapter sets out to describe the features of moving water, especially those which are important to the living organisms of streams.

Water accumulating in a stream or river originates as the result of preciptation of moisture from the atmosphere in the form of rain or snow. By no means all of this water actually reaches a stream. A fraction, large or small depending upon temperature and other factors, will be immediately evaporated back into the air. The water which actually reaches the ground may take various courses. Some is directly absorbed by plant roots and travels up the plant in the transpiration current to be evaporated into the atmosphere. Some enters the soil and percolates downwards until it reaches the water table formed by an impervious layer. This water will run downhill and probably never reach the stream but goes on to feed a river and eventually enters the sea. Some is surface water while some percolates to sub-surface layers, never arriving at the water table but reaching the stream by travelling through the soil. These different routes together contribute to the run-off entering a stream direct. Through the courses illustrated in Fig. 7.1 water enters a stream and collects soluble minerals and organic substances en route. In this way pollutants also enter a stream or river.

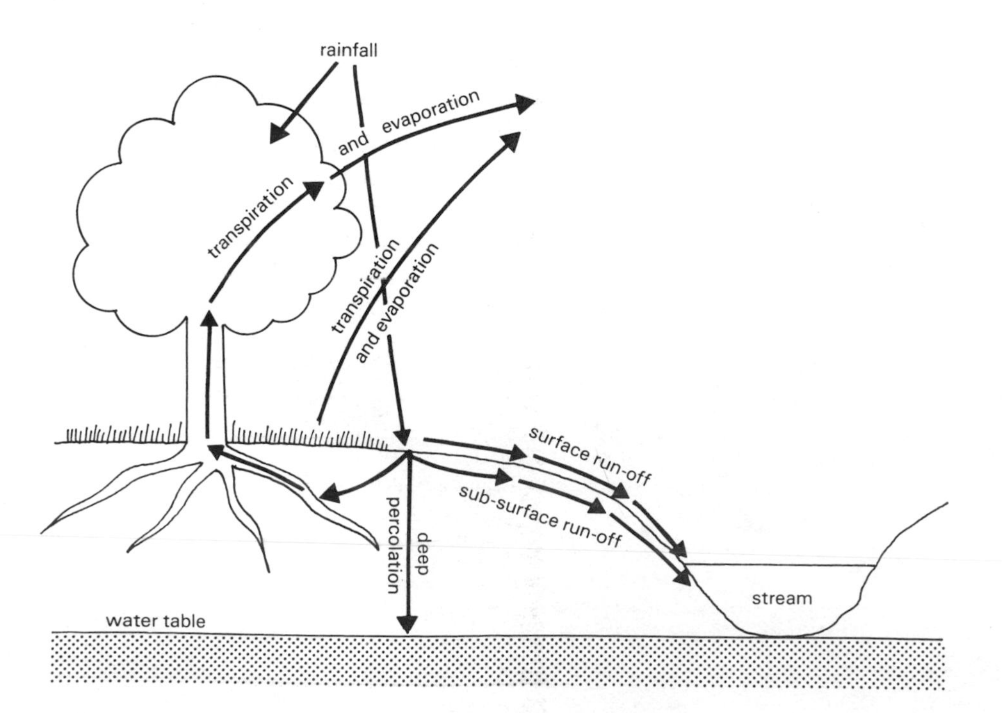

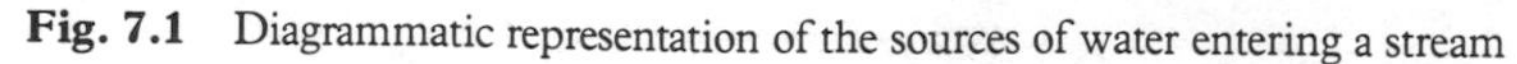
Fig. 7.1 Diagrammatic representation of the sources of water entering a stream

Water flow and the creation of microhabitats

As a stream flows downhill it becomes larger and collects water from an increasing area of catchment. The volume of water flowing in a stream is known as its **discharge** *(D)*. This is related to the cross-sectional area of the stream *(A)*. The relationship between *D* and *A* can be expressed as:

$$D = AV, \text{ where } V = \text{velocity.}$$

From this it follows that the velocity of a stream need not increase even if its discharge and cross-sectional area do, as when a stream becomes wider and carries more water.

Measurement of rate of flow in different parts of a stream can be made by using a variety of methods. Of particular interest here are those which can measure flow as near as possible to where animals and plants live (see Appendix 7C).

Stream flow is a complex matter, however, and a single stretch will include areas of water moving at different speeds. Measurements show that a current decreases as the bottom of a stream is approached for here friction will retard flow. This is of importance to the benthic fauna for the **boundary layer** of water flowing over a boulder in the bed of a stream may be only 1–3 mm thick, this layer becoming thinner as the flow increases. Indeed, as water flows over a boulder there is a zone of dead water on the downward side, and similar dead zones are created by other large obstructions such as tree trunks lying in mid-stream (Fig 7.2 (a)). These dead zones become microhabitats for certain benthic organisms capable of clinging to the upper surface of a boulder in the thin boundary layer.

A stony bottom in very swift, shallow reaches of a stream leads to riffles and the incorporation of bubbles of air providing a state of supersaturation. In the same way water falling over a waterfall (Plate 7.1) causes air to be trapped, a standing wave being formed a little distance below the fall. Air bubbles in the water reduce its density and thus the swimming thrust of fish such as salmon endeavouring to leap a waterfall. Thus successful individuals tend to take off from the standing wave (Fig. 7.2 (b)).

Water flowing downstream is perpetually shifting stones, pebbles and sand on the stream bed. Greatly increased flow after torrential rain can result in large boulders being shifted downstream. Table 7.1 shows how an increase in flow relates to the size of particles moved, although it should be stressed that these figures are only an approximation. Generally speaking, silt and mud particles settle out at a current speed of up to 20 cm s^{-1}. Larger mineral particles such as gravel and stones remain. Large stones offer shelter to smaller ones and to sand and gravel, each creating a microhabitat for the smaller fauna.

Table 7.1 The speeds of flow required to move mineral particles of different sizes (after Nielsen, 1950)

Speed of flow (cm s^{-1})	*Diameter of mineral particles moved* (cm)
10	0·3
25	1·3
50	5·0
75	11·0
100	20·0
150	45·0
200	80·0
300	180·0

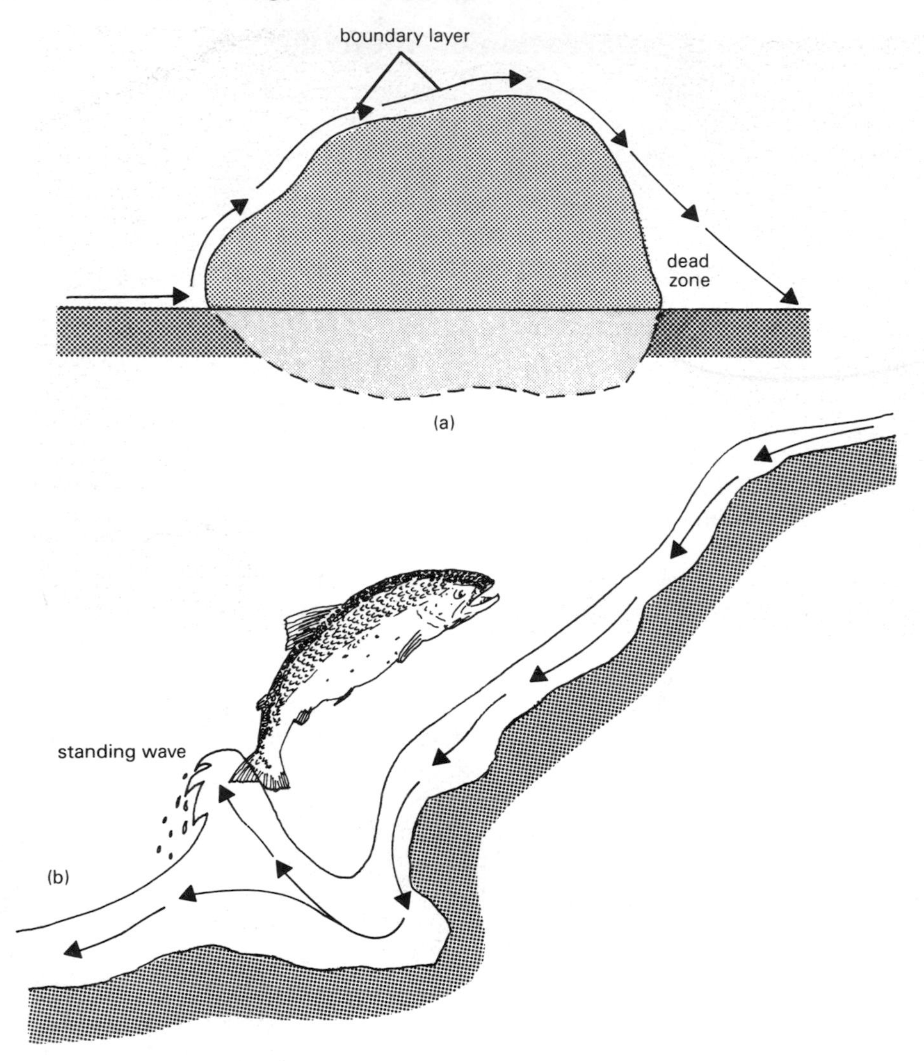

Fig. 7.2 (a) Flow of water over a boulder, showing boundary layer and dead zone. (b) Water flowing over a waterfall, showing standing wave below

Stream banks and patterns of flow

Even in the straight stretches of a stream the deeper regions often meander from one side to another, a tendency which we cannot completely explain. In negotiating a curve the main force of the current is towards the concave bank where it creates a deeper channel. The rush of water erodes the bank which may become undercut, the products of erosion being deposited further downstream on the same bank at the spot where the main stream sways across to the opposite bank (Fig. 7.3). A fallen log or any other large obstruction can cause a change in the direction of the current which, by erosion of the banks, can alter the stream courses. The important thing about the configuration of a stream is that it offers various habitats such as deeper water, shallow water, sand, silt, and organic debris, each supporting a different population of organisms. A greater variety of species is thus usually to be found in curving stretches than in straight ones.

Plate 7.1 Standing waves beneath waterfall, Morar Falls, Inverness-shire

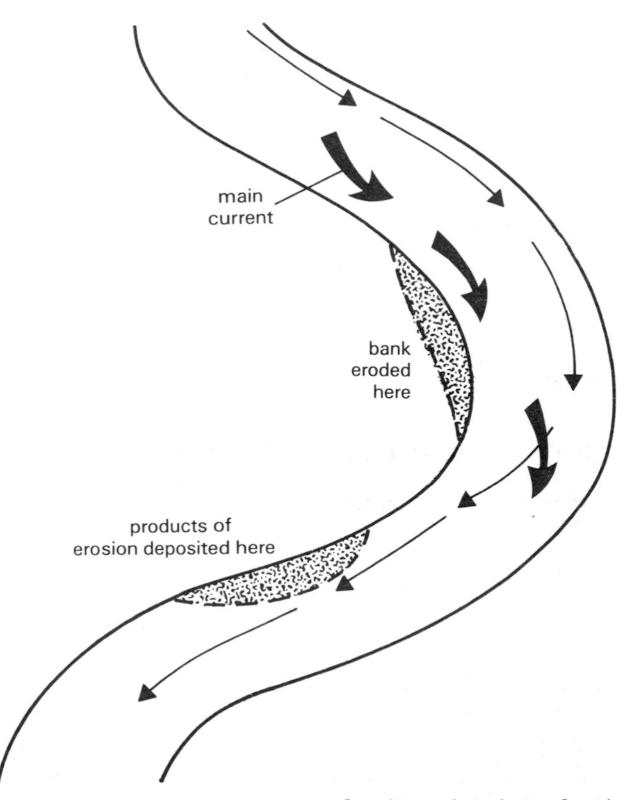

Fig. 7.3 Meandering stream showing areas of erosion and products of erosion

Since the biology of organisms living in the different regions of a stream is closely related to the nature of the stream bed, i.e. to the size of particles and the amounts of organic debris present, it is important to devise means of recording these factors more precisely. For methods of recording particle size, categories based on the Wentworth Scale, and estimating the organic content of samples, reference should be made to Cummins [7.1].

For many stream inhabitants a degree of shade is necessary. Bankside trees are therefore important, as well as overhanging plants and surface weeds in slower flowing reaches. Canals, such as the Grand Western Canal in Devon (Plate 7.2), can become densely covered by late summer with water lily leaves. While offering shelter, shade and spawning grounds to a number of fish and invertebrates, they can also prevent free gaseous exchange at the surface between air and water. Some clearance of weed may be necessary but should be done with care, avoiding clearance from bank to bank at any one time.

Some physical and chemical characteristics of flowing water

Temperature

The temperature of lotic water varies more rapidly than that of lakes but occurs over a smaller range, at least in the shallower parts of lakes.

Various factors cause changes. For instance, sunshine after heavy rainfall can result in a rise of temperature, possibly caused by water draining into a stream from warmed soil.

Streams are often spring-fed at their source and therefore receive cold water, which in summer is gradually warmed by the sun and by conduction from the ground as it flows down a valley. The rate of warming of the water slowly decreases downstream until a mean ambient temperature is reached. The reverse happens in the winter when a spring-fed stream may start relatively warm due to water coming from the deeper regions in the soil which has not cooled down to winter surface temperatures. Such water becomes cooler as it flows away from the source. It is interesting, therefore, to note that spring-fed streams in winter can be warmer

Plate 7.2 Water lilies on the Grand Western Canal in June

than those receiving water by run-off. Rivers, on the other hand, because of their turbulent flow, show less variance in temperature from their stream headwaters to their point of outflow.

Measurement of temperature at different points can be made simply by using a mercury thermometer attached firmly within a section of land drain and sunk in the bed of the stream. A series of readings can be taken in order to record both seasonal and diurnal variation in temperature at different points. Temperature will also vary with depth and an easier method of recording differences is the use of a resistance thermometer.

Light

As in static water, light is an important factor and the amount of light reaching different depths in a stream or river depends upon a number of variables. Flowing water is usually more turbid than static due to the amount of silt and debris present which increases enormously in times of flood. Riffles also cause occlusion of light rays. There will also be seasonal variation in the degree of light penetration, due in part to differences of temperature, which cause an alteration in the viscosity of water, causing silt to sink twice as fast at 23°C as it does at 0°C.

Shading of the surface, either by bankside trees or by overhanging plants and surface weeds, will reduce the penetration of light rays.

Chemical changes

In this brief account we are not concerned to describe in detail the various chemical changes that can take place in a stream. Nevertheless, the mineral content of running water varies greatly from one region to another according to the local climate and geology.

Rain is not pure water but contains dissolved gases, principally nitrogen, oxygen, and carbon dioxide, from the atmosphere as well as appreciable amounts of salts of sodium, potassium, chlorine, calcium, magnesium, and sulphur to mention a few. But because of carbonic acid (from dissolved carbon dioxide) and weak sulphuric acid (from sulphides) which occur in only minute amounts, the pH of rain is fairly low. As it passes through the soil, acids become neutralized. Streams arising from peaty areas containing a few minerals are usually acid and, because of the dissolved organic matter present, they can be very brown in colour.

Vegetation promotes the slow percolation of water. Thus the longer the time taken for run-off water to reach a stream, the more organic and inorganic matter will be carried into the stream.

It is not often realized that sea spray can be carried far inland, contributing various ions, notably sodium, magnesium, and chloride, to the soil from which the water drains.

As with large weed beds in static water, the photosynthetic activities of stream plants can lead to supersaturation with oxygen by day and, conversely, respiration at night leads to conditions of deoxygenation. But apart from these local changes, dissolved oxygen and carbon dioxide are usually in equilibrium with the atmosphere and so remain almost constant.

In small, swift-flowing streams the oxygen content is usually near or even above saturation but there can be variations between source and mouth, both diurnally and seasonally, due to a number of factors, such as temperature and local discharges, as well as to the activities of living organisms. Various conditions, notably the presence of chemical pollutants of one kind or another, can also cause a reduction in oxygen content, or even the deposition of a thick layer of decaying vegetable matter, both of which result in an increase in oxygen demand by decomposing organisms.

Carbon dioxide, if it occurs in fairly high concentrations, is usually lost rapidly to the atmosphere or combined as carbonates. The rate at which equilibrium between the dissolved oxygen and carbon dioxide of the water and that of air is restored, depends greatly upon the degree of turbulence, the roughness of the stream bed, and the presence along its length of weirs and waterfalls, all of which restore oxygen to the water. These factors all contribute to a more stable environment and to a swifter restitution of any disturbance in the amounts of dissolved gases.

As water flows from its source it will gain oxygen, particularly during the day, due to plant photosynthesis. If the source is a limestone spring or calcareous rock it will contain a high amount of calcium carbonate and will quickly lose carbon dioxide to the air with a consequent rise in pH. On the other hand a stream whose source is boggy land or acid swamp will also lose carbon dioxide, resulting in a rise in pH, although to a lesser degree. Such streams may contain iron and, as they flow and the dissolved oxygen is increased, the iron will be deposited on the surounding medium as ferric salts which become coated with a rust-coloured mass of iron bacteria.

Narrow, slow-flowing streams show greater fluctuation in carbon dioxide and oxygen content, and therefore of pH, than wide, swift stretches since slow-flowing water increases weed growth with consequent high levels of oxygen by day. The bottom sediments containing bacteria may also create an oxygen demand enhanced at night by the respiration of rooted plants.

Organic matter in streams

There are various ways in which organic matter enters a stream. Its constitution and amount will depend upon local conditions. For instance, where trees overhang the stream there will be a direct fall of litter which will include not only the leaves but bud scales, flowers, fruit and also the cast skins and droppings of leaf-dwelling insects. Although tree litter will be greater in autumn there is a continual contribution throughout the year and this will vary according to the tree species and their inhabitants. Apart from trees, other overhanging vegetation and its inhabitants all contribute to the organic litter, not to mention the excretions of waterfowl, fish, and aquatic invertebrates.

Some of the litter is washed downstream to accumulate in eddypools lower down, but a large proportion of the soluble substances contained in litter is quickly leached out, the process of decay being more rapid in leaves of deciduous trees than those of conifers. The litter quickly becomes colonized by micro-organisms — in the first place by fungi. These feed saprophytically upon the litter, incorporating nitrogen from the water and creating a rich food for invertebrate organisms.

It is clear that many of the chemical changes described here are the result of biological processes and highlight the fact that in the absence of living organisms, a stretch of flowing water would be a very uninteresting place.

Fieldwork

Identify two or three micro-habitats in a stream with a gravel bottom, such as upstream and downstream of a larger boulder, in a slow-flowing area and in a swifter area.

Using one of the methods given in the Appendices to this chapter, measure the velocity. Collect the invertebrates present at these sites. Is there any difference between the inhabitants of the swifter sites and those living in the areas of slower flow?

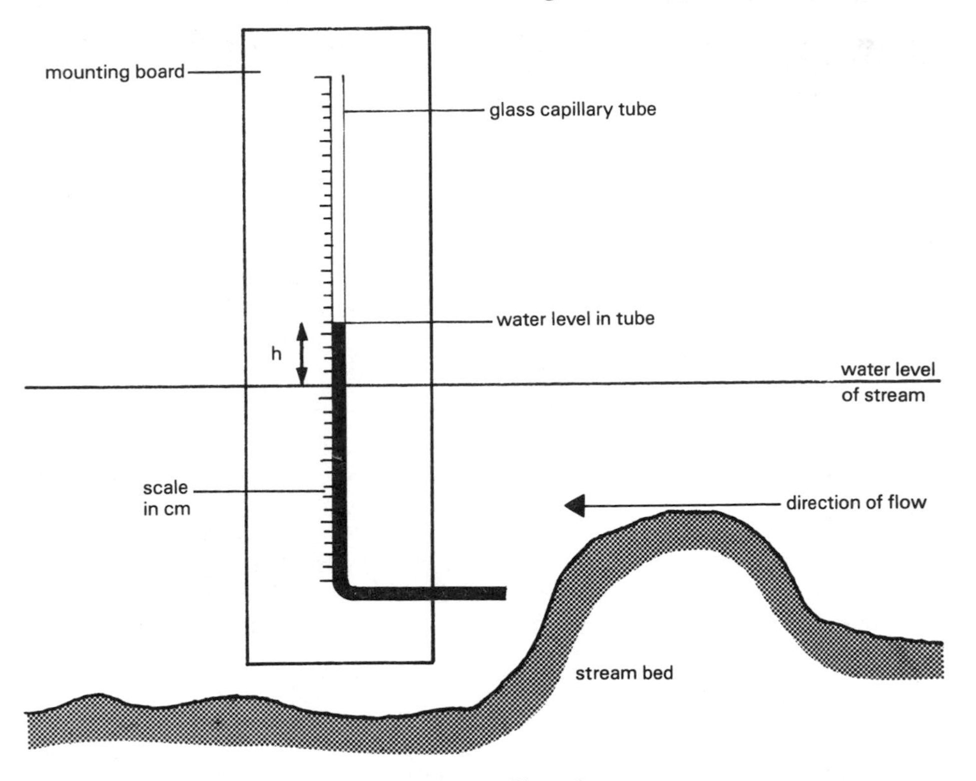

Fig. 7.4 Pitot tube

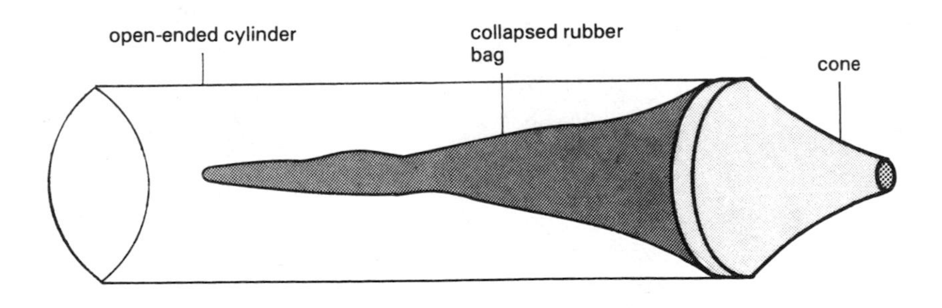

Fig. 7.5 A simple rubber-bag current meter

References

7.1 Cummins, K.W. (1962) 'An evaluation of some techniques for the collection and analysis of benthic samples with special emphasis on lotic waters', *Am. Midl. Nat.* **67**, 477–504

7.2 Nielsen, A, (1950) 'The torrential invertebrate fauna', *Oikos*, 177–96

Appendices: Methods of measuring velocity

Appendix 7A

Floats

This method involves timing the passage of a float such as a table tennis ball over a known distance. The fundamental problem with any type of float is that, when used for measuring the surface velocity, it produces a figure in excess of the true mean velocity of the stream. This can be overcome by multiplying the result by 0.8. Velocity also varies with depth. However, by repeating measurements over a 10 m stretch of water and taking the mean of time taken, the result can be expressed as m s^{-1} and used to compare velocity in different sections of a stream. Another method is to use coloured puffed rice to test areas of surface flow, but they will sink or be caught up over stretches of more than 100 m or so.

Appendix 7B

Pitot tube

This is an L-shaped glass tube of at least 5 mm diameter (Fig. 7.4). It is lowered into the stream with the short horizontal arm facing upstream. The velocity of the stream forces water up the tube and above the level of the surrounding water. The velocity of the stream is then calculated using the following formulae:

$$U^2 = h \times 2g$$

where U = velocity in m s^{-1}
g = acceleration due to gravity (usually 9.81)
h = height in m of water in tube above surrounding water.

Thus if height is 0.05m $U^2 = 0.05 \times 2 \times 9.81$

therefore $U = 0.99$ m s^{-1}.

However, this method has its problems for at very slow flow rates it is difficult to measure the height of the water column and at high flow rates a standing wave forms outside the tube, making it difficult to read the lowest measurement.

One major objection to this form of pitot tube is that the apparatus is too cumbersome to measure the flow rate at points close to the organisms. The following very simple method is worth description because it probably gives more reliable results than more complex and expensive micro-meters.

Appendix 7C

Rubber bag current meter

The apparatus consists of a truncated cone with an aperture of 5–10 mm in diameter at its apex. A thin rubber bag is attached to the base of the cone, protected by an open-ended cylinder (Fig. 7.5)

With the bag completely collapsed, the opening of the cone is closed with a thumb and the apparatus is placed, facing the current, at the point where the measurement is to be made. the thumb is now removed for a few seconds and then replaced. The amount of water collected is measured and from the area of the hole and the time of exposure the rate of flow is calculated. The measurements should be repeated 10 or 20 times at the same point to obtain consistent average results.

8 Plant communities of flowing water

At first glance it might be said that compared with the rich variety of aquatic plants found in still waters, which have been described in Chapter 5, those associated with flowing water are not so diverse. On closer investigation, however, we find that there is a large number of smaller species which colonize the stones of a stream bed, rocks and waterfalls, as well as the more obvious larger, rooted plants which often form extensive colonies. Flowing water can also support many of the same species found in still water in such places as eddy pools. The rich soil forming the banks of streams and rivers is colonized by other species associated with still water which often occur in great profusion.

Of the smaller plants algae are the most important but there are others, including some interesting bryophytes (mosses and liverworts) and stoneworts.

Stream algae and their colonization

The attached algae of running water grow on mud, on all kinds of rock and introduced debris, such as blocks of concrete and brick, as well as upon other plants. Some, like the frog-spawn alga, *Batrachospermum* (Fig. 8.1(a)), are quite large and conspicuous. This is one of the red algae and by reason of its pigments can tolerate shade conditions quite well. Attached algae must be firmly anchored to maintain themselves in a current. Some have a flattened thallus which grows close to the rock surface. Others, like *Vaucheria*, form cushions which divert the current. *Cladophora* (Fig. 8.1 (b)) and *Ulothrix* form long dense strings which in *Cladophora* can exceed 1 m.

Stream algae are often difficult to identify. For one thing, they have to be examined closely and dislodging them often breaks up the plants. By submerging glass slides in a stream comparative results can be obtained, in different current speeds and kinds of water, for the algae which colonize the glass. The first colonists are often diatoms, followed by larger species.

Factors controlling stream algae

Many of the species found in running water are opportunists, apparently able to flourish in a variety of conditions, the nature of which is not always apparent. At one time there may be a 'flush' of an apparently rare species which gives way later to colonization by others. Some of the green algae (Chlorophyceae), like *Cladophora* and *Oedogonium*, flourish in summer under conditions of higher water temperature and light intensity but are absent from shaded stretches of water.

Although many of the attached algae appear in both still and running waters, there are some which are only to be found in the latter. Of these species typical of flowing water, *Chaetophora* and *Draparnaldia* are common examples, both preferring to live in rapid water. *Batrachospermum*, on the other hand, prefers quiet, shaded currents. Current undoubtedly confers physiological richness on water because there is a constant renewal of materials in solution. This explains the demands of different species, some preferring a rapid flow of cool water with a high oxygen content and others colonizing warmer water with a lower content of dissolved gases.

Different substrates also affect colonization. On the whole, larger species of algae tend to grow on solid objects rather than on other plants. The red algae (Rhodophyceae) are slow-

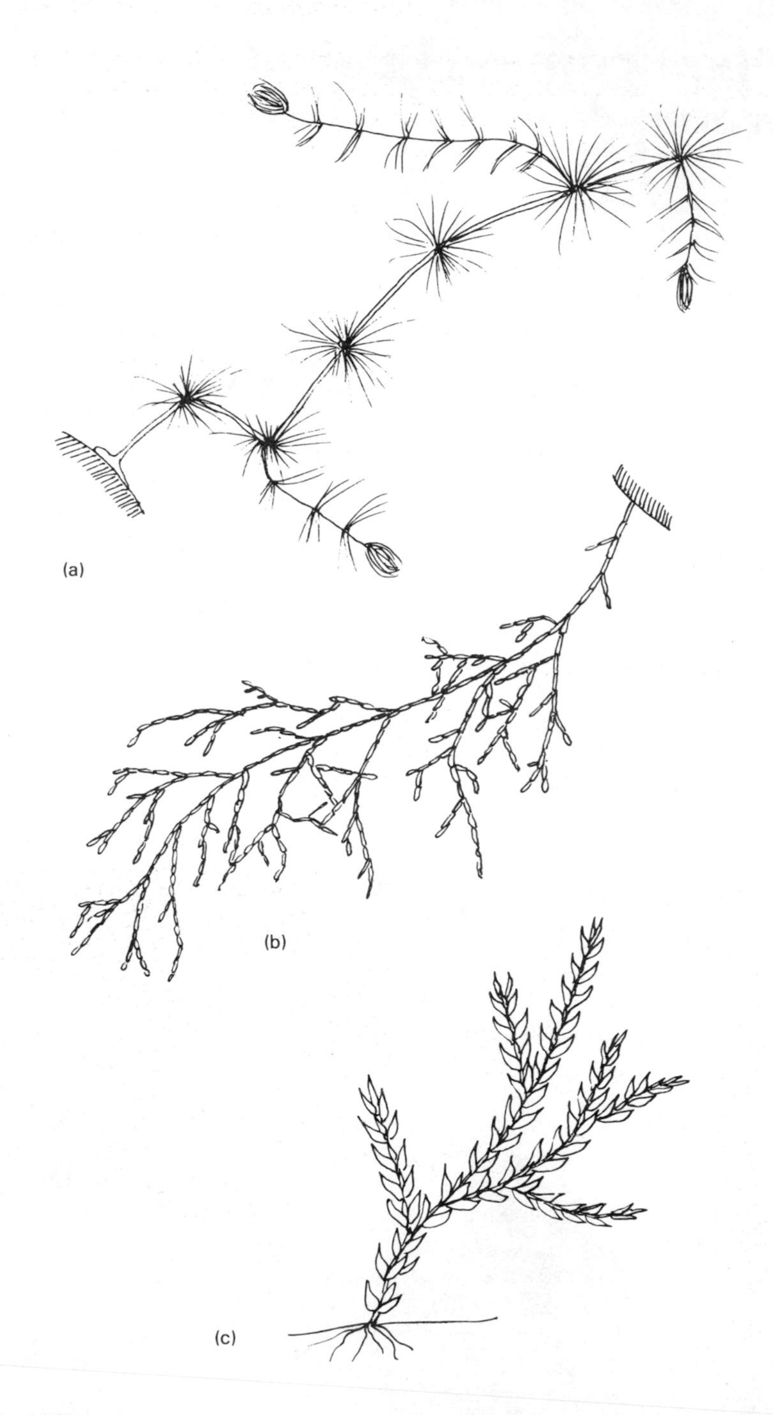

Fig. 8.1 (a) Frog-spawn alga, *Batrachospermum*. One of the red algae (Rhodophyceae). Visible to the naked eye as small, branched strings of beads composed of minute filaments. Colour usually greenish brown. Found attached to stones in shady places. (b) *Cladophora*. A green alga (Chlorophyceae) composed of long, branching filaments which form dense, dark-green masses attached to stones. (c) Water moss, *Fontinalis antipyretica* (Bryophyta). The stems, attached to stones, can reach a length of 40 cm or more in slow-flowing water. Forms dense masses on waterfalls

growing and are usually only found on rocks and large stones. Some of the larger plants can be colonized by a number of algae which may show preference for different parts of the same plant (Fig. 5.1(a)). It is an interesting fact that weeds with waxy leaves, like floating pondweed, *Potamogeton natans* (Plate 8.1), are not favoured by attached algae, possibly because of the slippery leaf surfaces.

The pH of a stream affects the algal species present. Many, such as *Cladophora*, are found in alkaline water, while *Batrachospermum* is equally at home in both soft, acid waters and hard limestone springs.

Since attached algae form the principal food of many grazing animals, it is not surprising that this is an important factor in causing patchy colonization. The majority of gastropod snails are algal grazers, and the caddis larva, *Agapetus fuscipes* (p. 78 and Fig. 9.5 (b)) can cause rapid depletion of algae on and around stones in a stream-bed. The grazing activities of the freshwater limpet, *Ancylastrum fluviatile* (p. 75 and Fig. 9.3 (b,c)), like its marine counterpart, are often obvious by the cleared patches around it.

Algae exhibit some degree of competition and succession in areas where conditions favour rapid growth of a species in one part of a stream, to be replaced by another species as the conditions change. Different species of attached algae can even colonize different parts of the same stone according to depth, different current speeds and so on.

The larger plants of running water

These comprise the lichens, mosses and liverworts, the stoneworts and the flowering plants. There are also the horsetails (Equisetales), some of which can withstand the mild flow of canals, although they are rarely found in stronger currents.

Plate 8.1 Floating pondweed, *Potamogeton natans.* Photographed in a silted backwater of the River Culm in Devon. Note the shining, floating leaves supporting the aerial flower spike

Like the larger plants of ponds and lakes, those of running water can be placed in distinct categories: some are rooted in the substratum, others attached to rocks or stones, while some float.

Among the floating weeds some of the duckweeds, such as *Lemna trisulca* and *L. minor*, occasionally get washed into a stream, becoming entangled with other plants. But they are not of great importance as colonists of British streams. Canadian pondweed, *Elodea canadensis*, which has been introduced into canals and waterways, spreads rapidly to form a thick mat on the surface, thereby interfering with water flow and causing deoxygenation of the waters beneath.

Attached plants are usually found on stones or rocks in swift water. One of the most common of these is the water moss, *Fontinalis antipyretica* (Fig. 8.1 (c)), which can form a dense growth on weirs and on stones beside waterfalls. It does not seem to be particular as to whether the water is soft, hard or even polluted. Since mosses cannot use bicarbonates in photosynthesis, they must have a ready supply of dissolved carbon dixoide. This is the reason why they grow more luxuriantly in turbid water which is enriched by the gas from the atmosphere. *Hypnum fluitans* is another fairly ubiquitous species of moss found on water-splashed stones and beneath the surface. Various kinds of liverworts also encrust rocks, with their flattened thalli wetted by the splash water.

Little is known about the factors which determine the distribution of mosses and liverworts, although depth of water is important since some are found in the splash areas while others, such as *F. pyretica* and *H. fluitans*, are truly aquatic.

Form and adaptability

It has been shown that the rate of respiration and assimilation of aquatic angiosperms increase with the oxygen concentration and that if this decreases, the effects are to a large extent offset by water movement. It is strange, therefore, that there are relatively few species of higher plants found actively growing in the main channel of a stream or river compared with the number of still-water forms. Also, although some rooted species, such as the river crowfoot, *Ranunculus fluitans*, and opposite-leaved pondweed, *Groenlandia densa*, are usually found in flowing water, there are none which are confined to this habitat, in contrast to algae and mosses of which there are many species exclusive to flowing water.

The buffeting produced by the current is often growth-inhibiting so that many plants have shorter internodes and smaller leaves. In *R. fluitans* the river forms have thread-like leaves on very long, flexible stems which trail downstream along the surface where there is good light intensity (Plate 8.2 (a,b)).

G. densa and *R. fluitans* are often found in still waters which favour lush growth and larger leaves. Silted streams offer conditions more closely resembling those of still water and here we can find plants such as wild celery, *Apium graveolens*, water-parsnip, *Sium latifolium* fool's watercress, *Apium nodiflorum* and more rarely, water lobelia, *Lobelia dortmanna*, which bears slender flowering spikes of mauve flowers. The strong aerial stems of the bur-reeds, *Sparganium* spp. (Fig. 8.2 (a)), can form dense strands in shallow water. Their prolific rooting systems collect silt, creating bog conditions and giving rise to the first stages of a hydrosere (p. 27).

Factors affecting distribution

The degree of hardness or softness of water is one of the major factors affecting distribution. Alternate-flowered water milfoil, *Myriophyllum alterniflorum*, occurs only in soft waters

(a)

(b)

Plate 8.2 (a, b) River crowfoot, *Ranunculus fluitans.* There is a number of closely-related species which are difficult to distinguish. The finely-divided leaves on long, stout stems float downstream, swaying to and fro in swift currents. The large white flowers appear in June and July

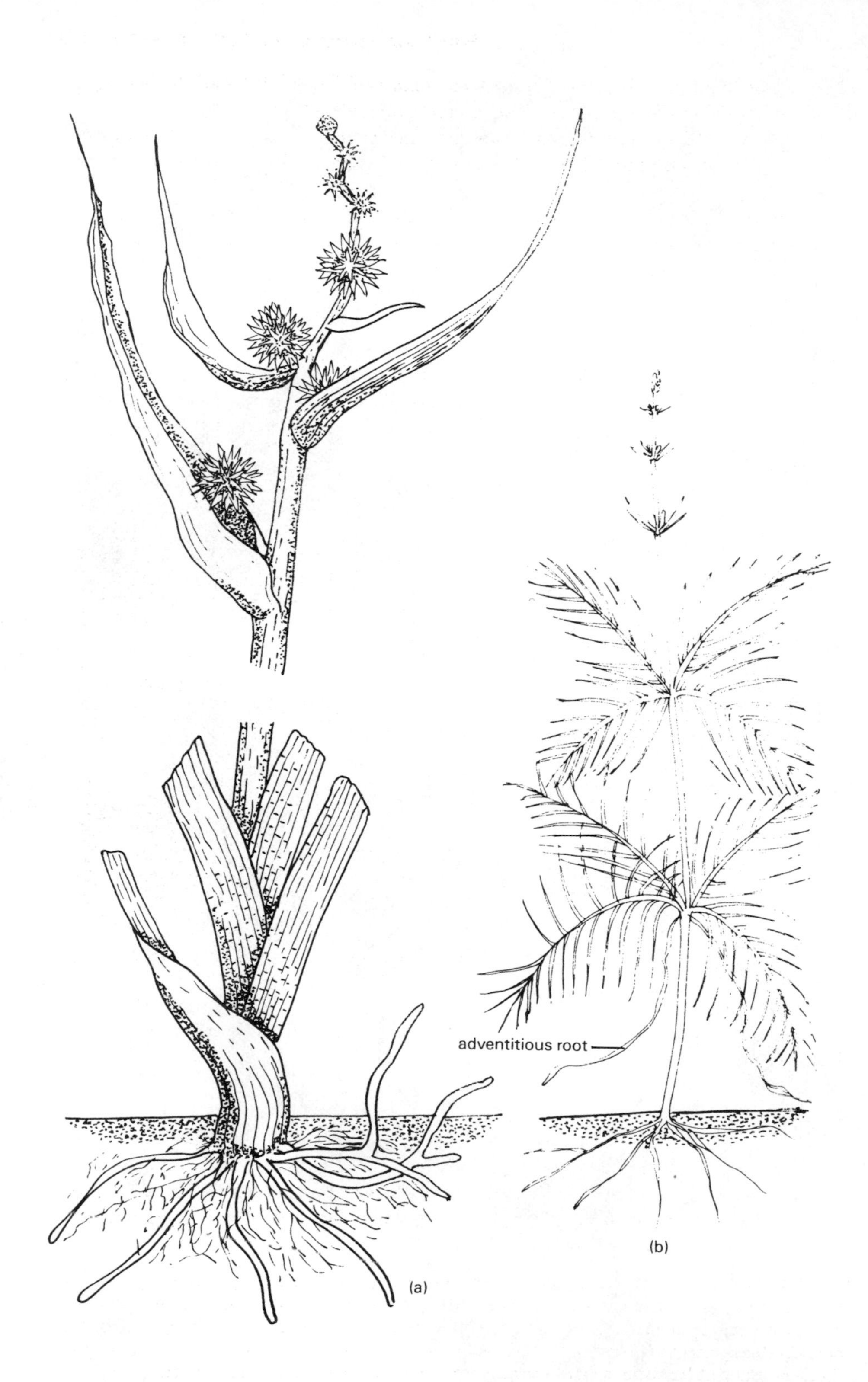
adventitious root
(a)
(b)

while its close relative, spiked water milfoil, *M. spicatum* (Fig. 8.2 (b)), requires hard water. *R. fluitans* appears to be equally at home in both hard and soft water.

The stoneworts, *Chara* spp. (Fig. 5.2 (b,c)) and *Nitella*, are typical of hard waters requiring a high calcium concentration. Their whorled leaves are usually covered with a thick deposit of lime.

The angiosperms of running water, like their still-water counterparts, do not flourish in the deeper parts of rivers due to the poor light intensity. Likewise, few are tolerant of deep, shaded areas. Hornwort, *Ceratophyllum demersum*, is one of the exceptions. Another is branched bur-reed, *Sparganium erectum* (Fig. 8.2 (a)), which thrives in such conditions. Depth is an important factor in the distribution of the reed, *Phragmites communis*, since its leaves are unable to photosynthesize under water. Thus it is usually restricted to areas where at least one third of the shoot is above water. The flowering rush, *Butomus umbellatus* (Plate 8.3), will apparently survive inundation by floods, a hazard not so easily withstood by other plants.

Plate 8.3 Flowering rush, *Butomus umbellatus.* Found locally in slow-flowing water of ditches or at the edge of lakes. The umbel of pink flowers is carried at the top of the flowering spike, surrounded by sharply-pointed linear leaves

Overwintering

As we have already seen, the larger plants colonizing flowing water need to anchor themselves firmly to the substratum. Most of them produce stolons and rhizomes by means of which they spread. In winter the stems and leaves of many of these plants die, leaving only

Fig. 8.2 (a) Branched bur-reed, *Sparganium erectum.* The thick rooting system and quickly-spreading tuberous roots soon build up dense stands of plants forming swamps and islands with other plants among them. (b) Spiked water milfoil, *Myriophyllum spicatum*, usually found in hard water. In this species the flower spikes are borne above the surface, supported by the submerged whorls of leaves. Adventitious roots, growing from the whorls, assist in spreading the plants

the rhizomes and stolons to survive. In spring these vegetative structures produce new growth. Some plants, however, persist through the winter unchanged save that growth stops and their leaves may be reduced. Table 8.1 lists some plants of both categories.

Table 8.1 Examples of persistent and non-persistent stream macrophytes in winter

Persistent in winter	*Dying down in winter*
Canadian pondweed, *Elodea canadensis*	Amphibious persicaria, *Polygonum amphibium*
Mud water starwort, *Callitriche stagnalis*	Arrowhead, *Sagittaria sagittifolia*
Opposite-leaved pondweed, *Groenlandia densa*	Broad-leaved pondweed, *Potamogeton natans*
River crowfoot, *Ranunculus fluitans*	Yellow water lily, *Nuphar lutea*
River water dropwort, *Oenanthe fluviatilis*	Water plantain, *Alisma plantago-aquatica*
Spiked water mifoil, *Myriophyllum spicatum*	
Fool's water cress, *Apium nodiflorum*	
Water cress, *Rorippa nasturtium-aquaticum*	

Roles of plants in flowing water

In the swift waters of weirs and waterfalls, dense growth of water mosses and liverworts can afford shelter to a number of invertebrates, notably several species of chironomid larvae (p. 78 and Fig. 9.4 (d)) which construct loose, silken tubes in the moss clumps and feed on the accumulated detritus. The larvae and pupae of the fly *Limnophora* sp. also use thick clumps of moss and algae to which they attach themselves by means of hooks.

The thick mat of rhizomes and stolons produced by larger plants in a stream bed afford shelter to many invertebrate species, ranging from encrusting sponges and bryozoans to aquatic flatworms and annelids, which make use of the silt trapped amongst the roots and the stores of oxygen in the hollow stems and leaves.

Emergent stems and leaves are important for caddises, mayflies, stoneflies, alderflies, and dragonflies, for the adults make use of these above-water structures for emergence from their pupal cases. Many of these insects use them also for laying their eggs, either just above or below the surface.

In swift currents or in time of flood and also under ordinary conditions, shoots and roots can break off and float downstream until they become attached to the bed, causing a local reduction of current. Downstream of an obstruction there will be a build-up of sediment onto which plants may spread. Eventually this results in a small bank and, later on, even a small island may be created which will alter the direction of the current (Fig. 8.3). Downstream of the island there may be scouring of the banks as a result of the changes higher up (Plate 8.4).

Fieldwork

1 Depth and shading often vary across a river, which affects the distribution of plants. Select several suitable sites along a river which offer differences of depth and shade and, using a Secchi disc (Chapter 2, p. 11), determine the depth of visibility across the river at different points at these sites. How does the distribution of the plants relate to your readings? Which plants prefer shade and which open water?
2 Make a list of the macrophytes present in a stream or river, noting for each species the method of perennation.

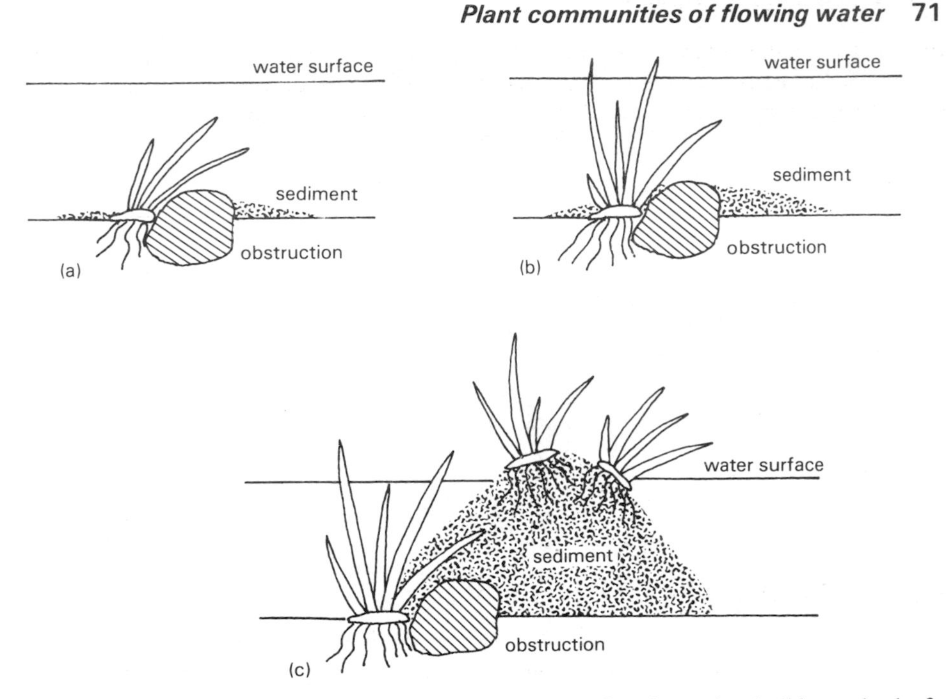

Fig. 8.3 Diagram to show how a plant taking root upstream of an obstruction, builds up a bank of sediment on the downstream side (a) and (b), and finally (c) the sediment builds up into an island colonized by other plants

Plate 8.4 The small island formed by the matted growth of rush and iris roots has caused a change in the direction of the current, resulting in the scouring of the far bank of the river

References

8.1 Whitford, L. A. (1960) 'The current effect and growth of fresh-water algae', *Trans. Am. Microsc. Soc.*, **79**, 302–9

9 Animal communities and movement in flowing water

As we saw in Chapter 7, flowing water offers diverse living conditions for invertebrates and fish. The current, which varies in speed along the length of a stream, causes the creation of numerous microhabitats in which different communities become established. The stream bed can vary from sand to silt, gravel, and large boulders, all of which offer changes in current speed and a variety of habitats. Sheltered areas with little current are inhabited by different populations from regions of swift current where the substratum is unstable. Here the requirements are for more specialized adaptations of structure and many of these modifications are different from those possessed by animals in still waters. A strong current demands modifications for clinging onto the substratum, for breathing and for movement, and only those animals able to adapt in these ways, either structurally or by their behaviour, can survive the rigorous conditions. Having said this, however, the successful species can be, and often are, present in large numbers. The habitats found in a weed-fringed canal or slow-flowing stream more nearly approximate to that of a pond and hence many organisms found there are common to both.

In this chapter we shall describe the various ways in which stream animals are adapted to their diverse environments. Any diversions adopted are, in a way, artificial, since some animals show several kinds of different structural modifications. These may be the result of response to several different factors.

Structural modifications

It is obvious that most stream organisms must be able to attach themselves in one way or another to stones or other objects in order to avoid being washed downstream.

Sessile animals are permanently attached. To this group belong the ciliated protozoans, freshwater sponges (Porifera), and the moss animalcules (Bryozoa).

Among the ciliated protozoans are the vorticellids, the old name for which was bell animalcules. Different species of *Vorticella* (Fig. 9.1 (a)) are sometimes found in dense masses attached to plants or the shells of molluscs and other animals. They are not colonial since each individual has a stalk, capable of rapid contraction, which is separately attached to the support. *Epistylis*, however, is a colonial ciliate (Fig. 9.1 (b)), all the animals being united on stiff branching stems attached to rough stones or tree roots.

Most sponges are marine and freshwater sponges are often difficult to find. They must be sought attached to fallen logs, tree roots or encrusting lockgates. The species usually encountered in streams or rivers is called the river sponge, *Spongilla fluviatilis* (Fig. 6.2 (a)). It is a colonial animal and because it is a filter feeder it cannot tolerate too much sedimentary matter which might clog the numerous feeding pores.

The moss animalcules (Bryozoa) often elude notice until one's eye is attuned to the search. One of the more commonly found species, *Plumatella repens*, is illustrated in Fig. 9.1 (c,d). It forms a greyish, branching colony, adhering closely to underwater objects, particularly tree roots and stones. By the end of the summer the colony may have expanded to cover an area of up to a square foot, each branch putting out retractile, horseshoe-shaped crowns of tentacles (lophophores) which trap microscopic organisms in the detritus floating by. These colonies are often the home of other organisms which benefit from the current of detritus circulating round the lophophores. Some species of *Vorticella* may be found attached to the tubes, and ostracods and small annelids pass in and out of them.

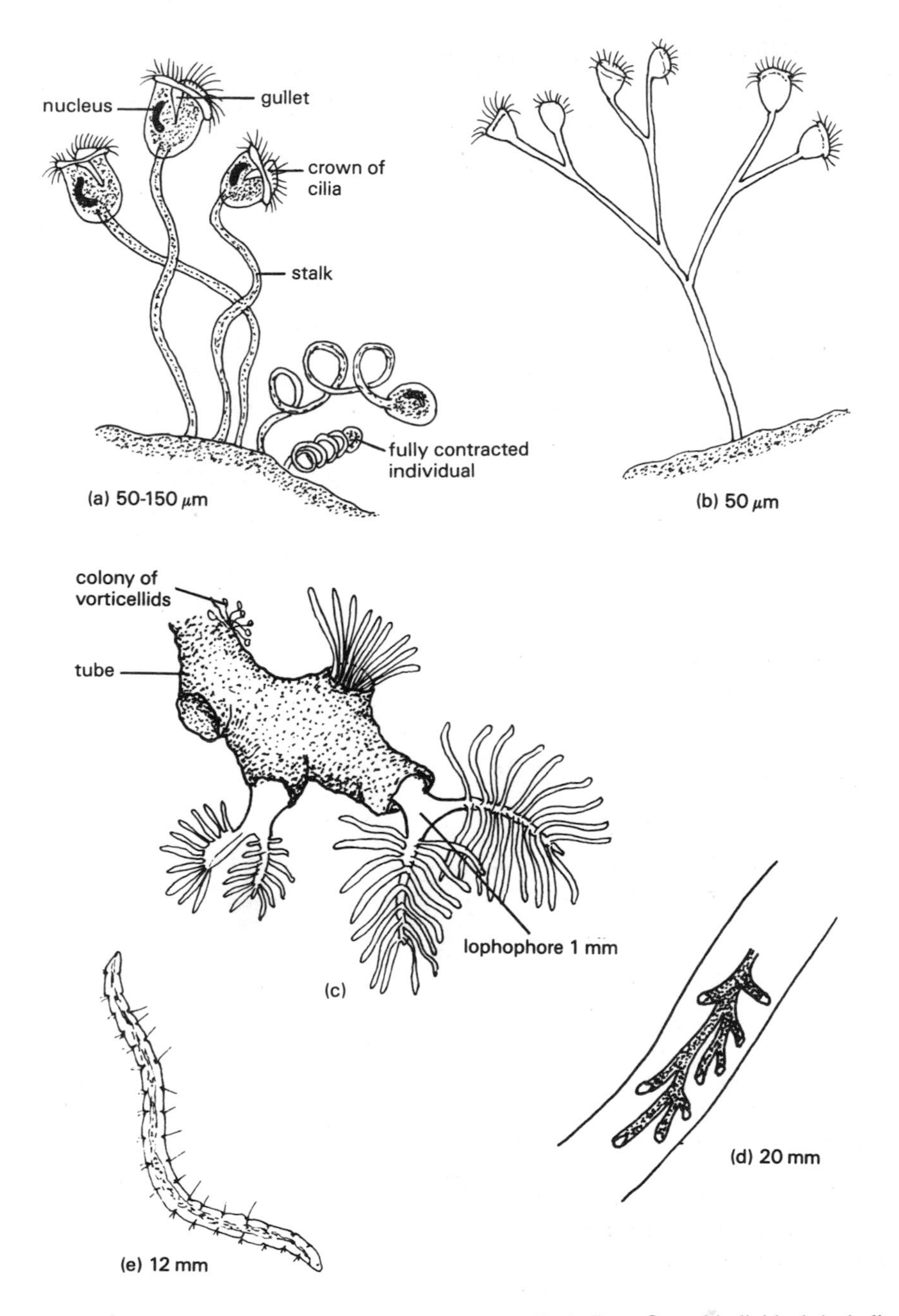

Fig. 9.1 Some sessile invertebrates of flowing water. (a) *Vorticella* sp. Several individuals including two in which the cilia are withdrawn and the stalks have contracted spirally. The crown of cilia waft particles of food into the gullet to become digested within the body. (b) Branching colony of *Epistylis* in which the stalks are non-contractile. (c) *Plumatella repens.* Part of a colony found adhering to a broken concrete block in the bed of a stream in East Devon. The contractile lophophores are extremely sensitive to vibration and to changes in light intensity when they contract within the tubes and may take a long time to extend once more. Note the colony of vorticellids attached to the tube. (d) Part of a colony of *P. repens* attached to an underwater object. They favour shady places, such as under canal bridges. (e) *Nais.* A small annelid which finds shelter within the tubes of *P. repens*

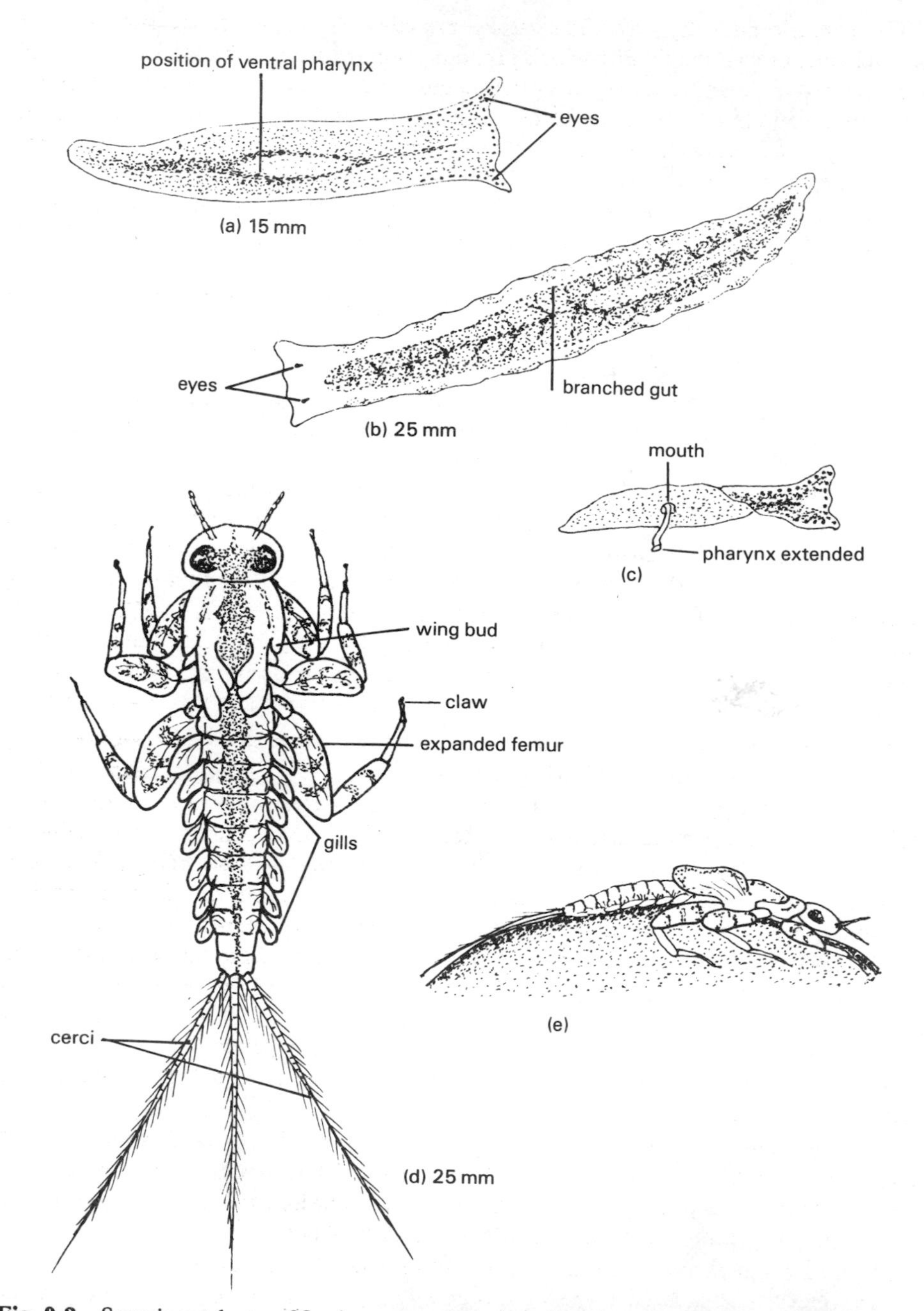

Fig. 9.2 Some invertebrates of flowing water which show fattening of the body. (a) *Polycelis felina* and (b) *Dendrocoelum lacteum*, two free-living flatworms (Platyhelminthes) frequently found under stones in flowing water. Note the position and number of eyes in each. (c) *P. felina* showing the muscular pharynx protruded through the mouth. Food is sucked up through the pharynx, digested within the branched gut and waste matter eliminated through the mouth. Food is located by sensory cells on the head region and large numbers of individuals can be attracted by small pieces of raw meat suspended in the water. (d) *Ecdyonurus* sp. (Ephemeroptera). The broad, flattened body of this mayfly nymph and the large flattened limbs, ending in hooked claws, maintain a hold on the upper surface of stones in swift currents (e)

These sessile animals, with their rather specialist requirements, are not common, probably due to their immobility which prevents them from migrating to more favourable positions when the stream is in spate or in times of drought when it may dry up altogether. It cannot be said that these sessile animals contribute significantly to the economy of the freshwater community. Nevertheless, by their ciliary feeding methods they are constantly sieving the water to extract microscopic organisms, thereby assisting in the cleaning process. Undoubtedly their chief role is to offer shelter to other organisms, but so far as man is concerned, large colonies of bryozoans can create a nuisance by causing blockage of water and sewage pipes.

Unattached stream animals must adopt one or other of several methods to prevent them from being easily dislodged.

Body flattening is usual and found in many animals that live on or under stones within the boundary layer. These are the **lithophilous** animals. Flattening of the body is, however, not confined to lithophilous species for it is also a feature of the nymphs of many species of mayflies and stoneflies which do not live exposed to the current.

Leeches and flatworms (Figs. 6.3 and 9.2) always have flattened bodies which enable them to shelter beneath stones. The surface of the body of a flatworm is covered with cilia, kept in constant motion, which enable it to glide over surfaces. At the same time, glands secrete slime which lubricates the surface and no doubt also assists the animal in maintaining position in a current. The nymphs of the mayfly, *Ecdyonurus* sp. (Fig. 9.2 (c,d)), live on the surface of stones, and the expanded joints and hooks on their legs assist in maintaining their position. Two common species of fish, the bullhead and stone loach (Fig. 10.11 and p. 100) found among stones on the stream bed, have flattened bodies which are also streamlined.

Streamlining of the body offers the least resistance to water and is also found among many stream invertebrates. The tapering bodies of the nymphs of the mayfly, *Baetis rhodani*, illustrate this well (Fig. 9.3 (a)), and are often among the most numerous animals in the stream bed. These nymphs actually stand fairly high on their legs, with the abdomen hanging free while the tail swings from side to side keeping the head facing the current. Fringes of hairs bordering the tail filaments, when moved up and down, form an efficient swimming organ, the tail bristles becoming folded at rest. In swift currents the whole body can be closely applied to a stone surface, enabling it to remain within the boundary layer.

The river limpet, *Ancylastrum fluviatile* (Fig. 9.3 (b,c)), is not only streamlined but also possesses a flattened foot by means of which, like its marine relatives, it can cling strongly to a stone. To a lesser degree the wandering pond snail, *Limnaea pereger* (Fig. 9.3 (d)), which is commonly found in flowing water, is also streamlined, and it, too, possesses a large flattened, muscular foot.

Body streamlining is also a feature of many of the cased caddis larvae. *Anabolia* (Fig. 9.3 (e)) adorns its tapering case with long pieces of twig and although quite clumsy, it clings by means of hooks on its legs which project from the case. The long twigs swing this way and that, helping it to keep facing the current as it clambers about.

Leeches are without doubt the best examples of animals which attach themselves by means of **suckers**. When the head of the leech is extended, the large posterior sucker holds the animal in position until the smaller anterior sucker gains a grip (Fig. 9.4 (a)). The posterior sucker is then released and the hind end is drawn up to meet the head in a looping movement. In this way, there is always one sucker holding the animal in place.

The larvae of the blackfly, *Simulium* (Fig. 9.4 (b)), are to be found in large numbers at certain seasons, attached to stones in the stream bed. These are some of the most highly adapted animals to life in rapid water. *Simulium* possesses large salivary glands which secrete a pad of sticky silk on the substratum to which it attaches itself by means of a pair of modified

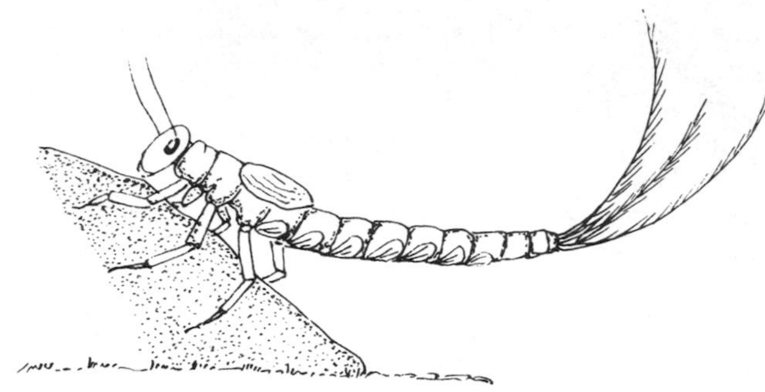

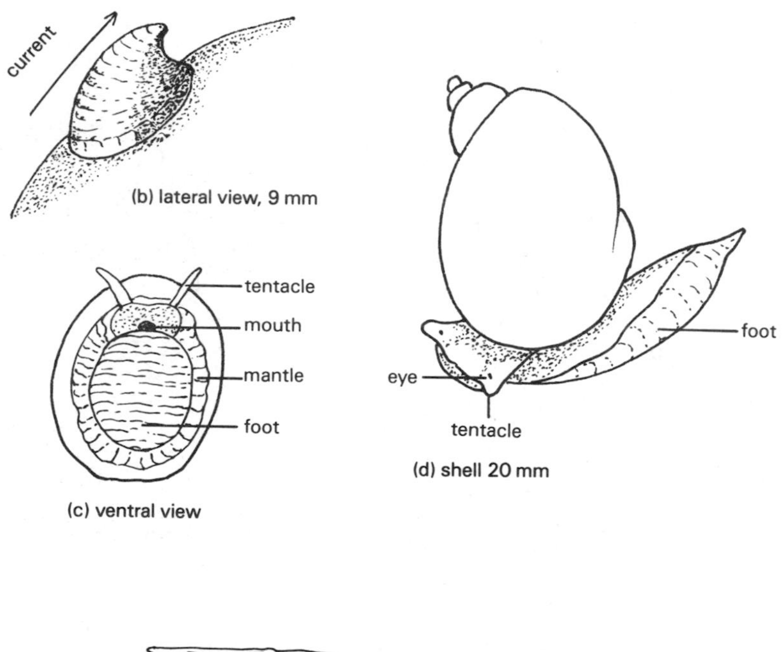

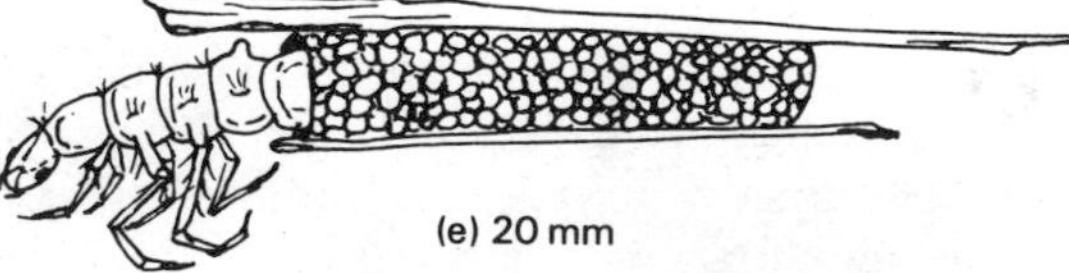

Fig. 9.3 Some invertebrates of flowing water which show streamlining of the body. (a) *Baetis rhodani.* The commonest of the mayflies. The nymphs cling, when at rest, to stones in the stream bed by means of strong claws. In swifter currents the body is lowered towards the substratum to lie within the boundary layer. (b) and (c) River limpet, *Ancylastrum fluviatile.* Pointed like a Phrygian cap, the shell of these small gastropods is curved to resist the current. Like marine limpets, they browse algae on stones, to which they cling with a large, circular muscular foot. (d) Wandering pond snail, *Limnaea pereger.* A common species in both still and flowing water. To some extent the whorled shell deflects the current and station is maintained by the large flattened foot. (e) *Anabolia nervosa.* A caddis larva which clings with its long, hooked legs to underwater objects. The case is composed of sand grains, to which it attaches long twigs that assist in keeping it facing the current

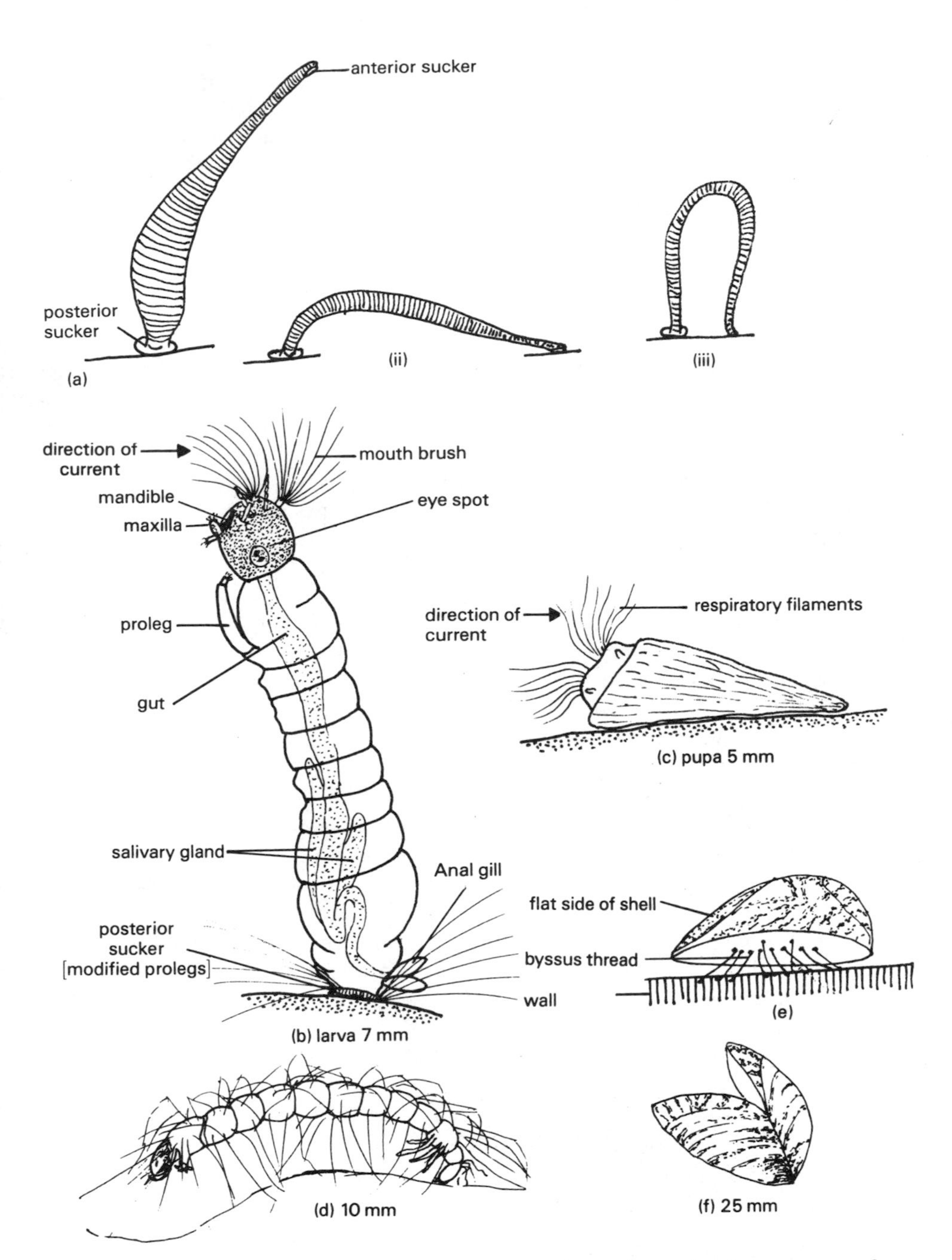

Fig. 9.4 Suckers, webs and glue as means of attachment. (a) Leeches use looping as a means of progression, always maintaining hold on the substratum by at least one of the two suckers each possesses. (b) and (c) The larvae and pupae of the blackfly, *Simulium* sp. always position themselves so that they face the current, the larva trapping food particles in the mouth brushes which form a fan on top of the head. (d) Larva of the midge, *Chironomus* sp. within a loose silken net which anchors it to the substratum. (e) Zebra mussel, *Dreissena polymorpha*, lateral view, showing the flattened side of the bivalve shell with byssus threads glueing it to stones or wooden piling. One from a large colony on the stone walls of the lock at the seaward end of the Exeter Canal. (f) Bivalve shell of zebra mussel showing distinctive markings

prolegs at the rear of the abdomen. Firmly attached in this way, the larva can hang free in the current with its mouth bristles facing upstream, ready to trap passing detritus. Moving from one place to another presents little difficulty as a fresh pad of silk is constructed and the front pair of modified prolegs become attached to the pad while the posterior pair are released, allowing the animal to progress in a leech-like manner. The funnel-shaped case of the pupa, directed upstream, is also firmly attached to the substratum (Fig. 9.4 (c)).

All cased caddis larvae attach their cases to some solid object when they pupate. This is also true of those which do not possess cases but construct rough pupation chambers using silk and small stones.

We must not forget the insignificant although often extremely numerous midge larvae, the chironomids, which form a large proportion of the benthic fauna of streams, inhabiting moss-covered rocks or underwater stems and roots of plants. There are a great many species and most construct loose cases of silken threads which attach them to the substratum (p. 70 and Fig. 9.4 (d)).

Lastly there is the zebra mussel, *Dreissena polymorpha* (Fig. 9.4 (e,f)), a comparatively recent introduction to Great Britain, probably on timber ships from the Baltic. Although not common, these mussels can be found locally in large numbers, attached permanently by byssus threads to lock gates and other man-made structures. Their strong means of attachment in an area where there is an abundant food supply may account for their locally large colonies. It is, however, unlikely that they could become established in swift-flowing water since their free-swimming veliger larvae would be swept away.

We have already mentioned **hooks** as a means of attachment in connection with caddis larvae. But nearly all larval and many adult freshwater insects have well-developed **claws**, by means of which they can maintain station under normal current conditions by holding onto rough stones.

A good example is the larva of the caddis, *Rhyacophila* (Fig. 9.5 (a)). This is a caseless larva living on the stream bed among stones and gravel, often in quite a fast current. Here it can move about quite rapidly, using the strong claws on the last abdominal segment which are brought forward beneath the thorax to grip the substratum. A number of other caddis larvae use their strong tarsal joints for gripping. In addition, many of the cased stream-dwelling caddises, such as *Agapetus fuscipes* (Fig. 9.5 (b,c)), use stones, often quite large ones, for the construction of their cases. They are therefore weighted and less easily dislodged.

The larva of the fly *Limnophora* (Fig. 9.5 (d)), is able to maintain itself in its favourite haunt, a waterfall, by the use of hooklike prolegs on the end of the abdomen which enable it to cling to thick growths of water moss and algae.

Movement and population changes

Maintaining position against the current is a constant challenge to animals living in a stream. Equally important is selection of the right speed of current to suit their needs. The numbers of invertebrates which in fact lose contact with the substratum and are carried downstream is quite considerable, for if a net is suspended above the bottom, facing upstream, it will soon collect an appreciable number of benthic animals.

The picture is complex for there are both upstream and downstream movements, also random small migrations along the stream bed or through the substratum. Not all such movements are influenced by current alone. Other factors are important and the reactions of the animals may be controlled by one or more of these.

It is difficult to subject stream invertebrates to laboratory experiments aimed at measuring the effects of one or other factor determining their movements. Variables such as exposure

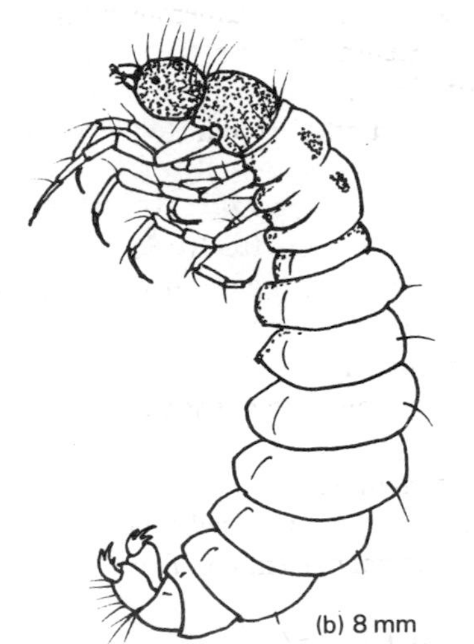

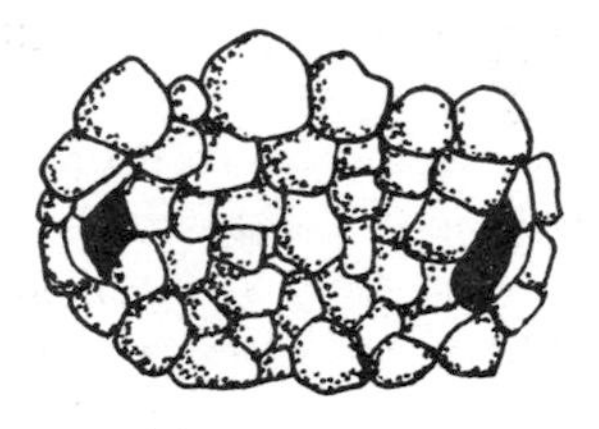

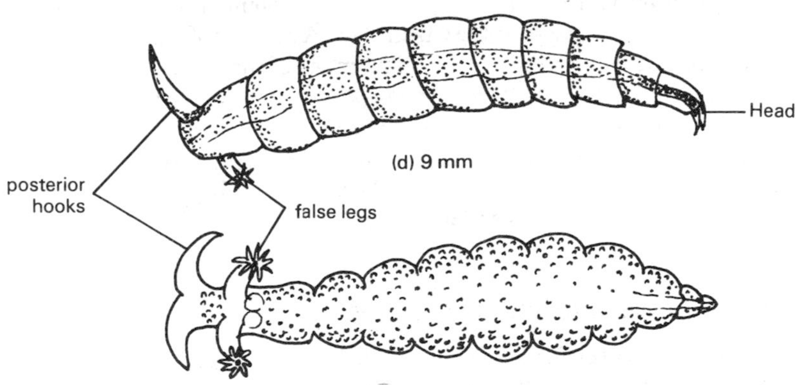

Fig. 9.5 Hooks and stones as a means of attachment. (a) Caseless caddis larva, *Rhyacophila dorsalis*, living in fast current in the gravel beds of streams and rivers. It moves about actively in search of prey in the gravel and, on pupation, builds a rough shelter of stones. (b) The larva of the caddis, *Agapetus fuscipes* makes a case of small and quite large stones (c) glued to boulders in the stream bed. Hooks on the last abdominal segment keep the larva within its case. Larva (d) and pupa (e) of the fly *Limnophora* sp. A pair of large hooks at the posterior end of the larva grip water moss and algae, keeping the larva in position. Hooks persist in the pupa, in which the skin is covered with small pimple-like projections

to light, availability and suitability of food, temperature, and the nature of the substratum may all influence their behaviour to some degree.

In order to investigate movement, colonization of new areas of a stream and population changes of stream animals under partially controlled physical and chemical conditions, Ladle [9.3] constructed an experimental recirculating stream channel. This was 53 m long with a trapezoidal cross-section, 1 m wide at the base and 2 m in width at the rim. The substratum simulated that of a chalk stream consisting of flint gravel 0.3–0.5 m deep. Such factors as variable water quality and changes in rate of flow could be controlled. Since macrophytes were absent in the channel, allochthonous input was negligible and a restricted invertebrate community was present. Upstream movement in the channel was minimal so that sampling of enclosed populations was possible.

The initial colonization of the substratum by invertebrates was chiefly the result of oviposition by insects [9.4]. Some arrived with the inflow water whose source was a bore-hole. The first colonists were the chironomids whose numbers rapidly increased as they grazed the abundant growth of diatoms on the gravel. During the months April–December records showed that in the absence of drift and upstream migration, both of which are involved in colonization of substrata in natural streams, recruitment of insect larvae was rapidly brought about by oviposition, chironomids and ephemeropterans being the most abundant.

Under natural conditions upstream movement is not easy to observe although results of such movements can be seen by the arrival of new populations. The comparatively recent spread of Jenkins' spire shell, *Potamopyrgus jenkinsi* (Fig. 9.6), originally an inhabitant of the saline water of estuaries, to become established in most fresh waterways, is evidence of upstream movement. Similar movements of other molluscs have also been observed and records show that the wandering pond snail, *Limnaea pereger*, can move some 2.4 km in a year following its elimination in one area by pollution. These molluscan migrations, and similar movements of the larger crustaceans, are comparatively easy to record. Nevertheless, upstream movement probably takes place in a large number of the smaller species which are less easy to observe. Indeed, if the upstream reaches of a stream are not to be denuded altogether of animal life due to current drift, it stands to reason that counter movements upstream must take place in order to replenish populations.

Elliott [9.1], working on a Dartmoor upland stream, sampled invertebrate drift by means of a surface net made of nylon sifting cloth, with a rectangular mouth of known dimensions.

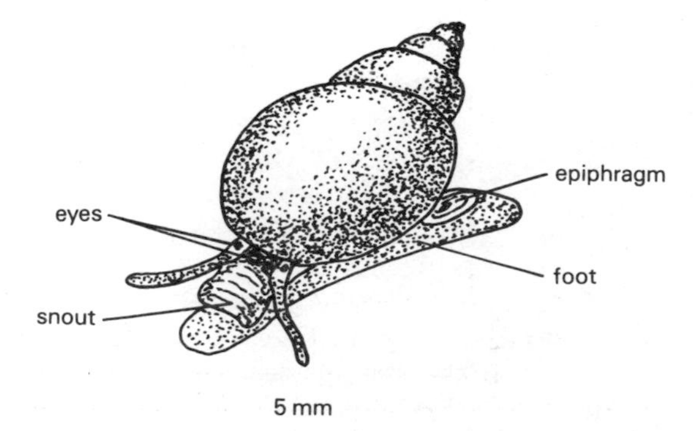

Fig. 9.6 Jenkins' spire shell, *Potamopyrgus jenkinsi.* Only during the last century has this small gastropod spread from estuaries to invade most British waterways. Its rapid spread is due possibly to two factors: it is hermaphrodite and appears to have no known parasites

This was attached by hooks to a metal frame at each end of which were floats (Fig. 9.7 (a)). The whole apparatus was moored to each bank of the stream by means of ropes and secured so that the bottom of the frame was 7 cm below the surface. In this way the volume of water passing through the net over a certain period of time could be calculated. Records were taken each month, when the nets were emptied every 3 h over a period of 24 h and the numbers of organisms entering during each 3 h period were noted.

In addition to the drift samplers placed at strategic points along the stream, several modified high-speed plankton samplers (Fig. 9.7 (b)) were also used. These consisted of a metal tube enclosing a nylon net of 15.5 meshes cm^{-1}. The size of mesh was sufficiently small to trap organisms such as small insect larvae, 1–2 mm long, without becoming clogged. A flow-meter, fixed at the rear end of the tube, recorded the volume of water passing through the net. These tubes were kept in place by iron rods attached to side brackets and driven into the stream bed.

There is a strong correlation between the quantity of aquatic drift and the amount of water flowing down a stream. Elliott's results (Table 9.1) show that, using surface nets, the amounts of water and drift in June, July and September, 1963 were both twice those of the corresponding months in 1964. Furthermore, at all seasons more aquatic invertebrates were found in drift samples taken at night than during the day, drift numbers probably being related to light intensity. Figure 9.8 shows results from drift samples, taken three-hourly during 24 h in March, April, and May. For most species taken in drift samples, the daily fluctuations in numbers followed that of the total numbers.

By taking samples also of the benthos, at sites below the surface nets, it was clear that the invertebrate drift was very local, most organisms being returned to the benthos after travelling only a short distance. Also, with the exception of triclads and trichopterous larvae with firmly attached cases, all species taken in bottom samples were also taken in the drift.

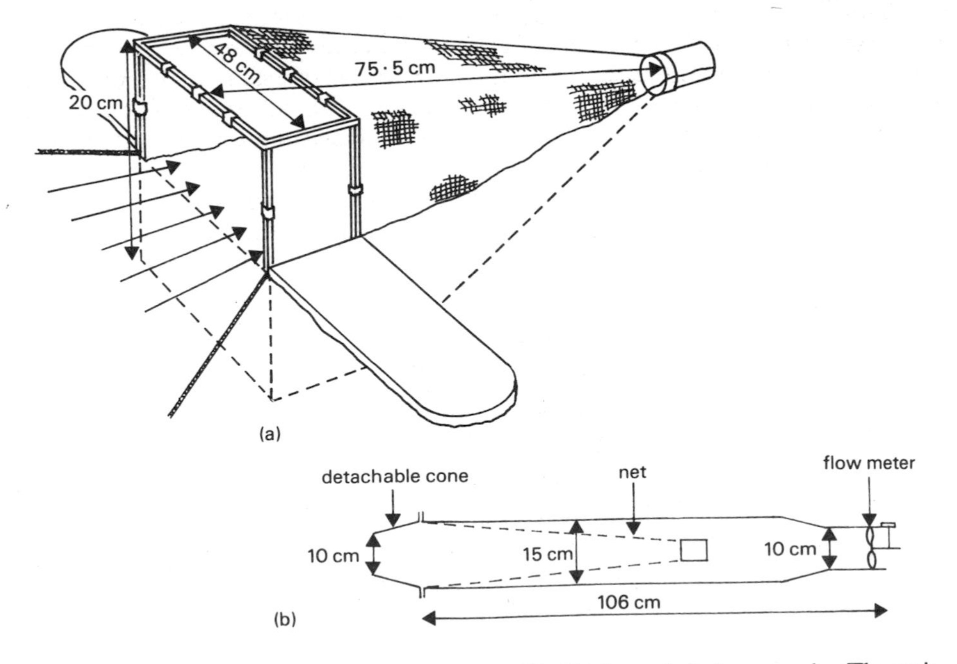

Fig. 9.7 (a) Surface net for sampling drift. (b) Modified high-speed plankton sampler. The net is held in place by a detachable cone which reduces the effective sampling aperture to an area of 78·5 cm^{-2} (after Elliott, 1967)

Table 9.1 Aquatic drift using surface nets, sampled from June to October in two successive years in Walla Brook, Dartmoor (selected from Elliott, 1967)

		June	July	Aug	Sept	Oct
Volume sampled	1963	2144	1560	2768	1600	1592
(1000 l/24h)	1964	1048	720	624	984	1572
Aquatic drift	1963	763	1036	1998	770	477
(number/24h)	1964	267	619	261	217	443

Flight is an important way in which upstream movement can be assured for winged insects. By placing nets upstream and downstream on a river, Roos [9.5] showed that, although emerging adults belonging to the orders Trichoptera (caddises), Ephemeroptera (mayflies), Plecoptera (stoneflies), and females of *Simulium* flew both up and downstream, movement was predominantly upstream, especially for ovipositioning females. In this way their eggs were distributed some distance upstream.

The significance of invertebrate drift

In extreme conditions of greatly increased flow due to flooding, there is a scouring out of the stream substratum, the material and the organisms it contains being removed and deposited

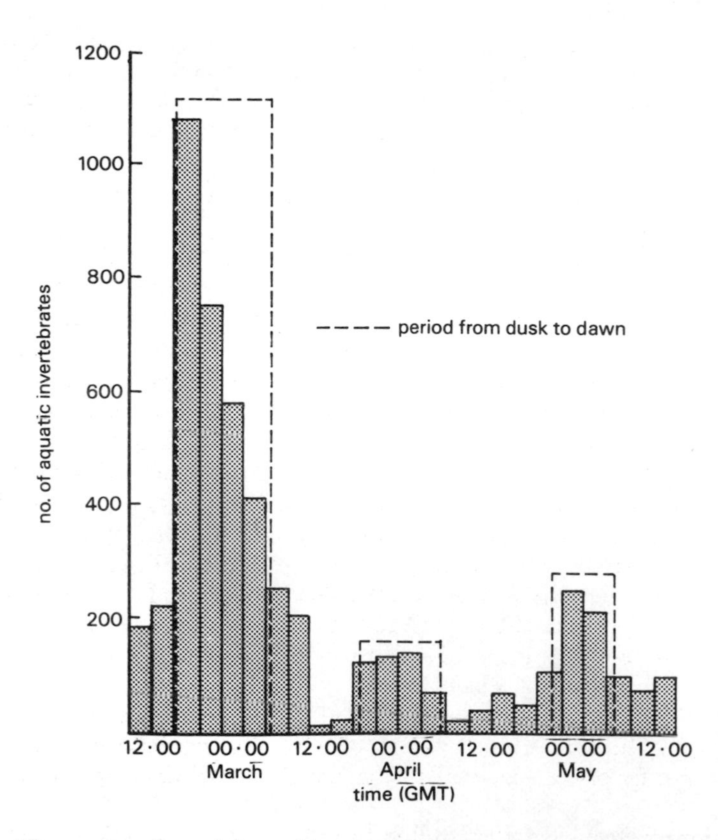

Fig. 9.8 The number of aquatic invertebrates in a Dartmoor stream taken in each sample for a period of 24 hours, in each of the months March to May, 1964 (redrawn from Elliott, 1967)

some distance downstream. Such movement, often resulting in the displacement of much of the benthic community, is called **catastrophic drift**. A more usual, and ever-present, movement downstream involves the displacement of benthic invertebrates which have lost their hold on the substratum to be carried by the current downstream. With slow flow there is constant movement of the benthos downstream.

After denudation of an area of stream bed, either by catastrophic drift or intentionally as a field experiment, it will soon become recolonized, often in the space of a week or two. This serves to demonstrate that a large-scale movement of organisms must be taking place all the time. As we have already mentioned (p. 80) there are several sources from which recolonization can take place: invertebrate drift, an upstream movement of organisms in contact with the substratum, an upward, vertical movement from the depths of the substratum to its surface, and from eggs laid by winged insects.

The fieldwork carried out by Elliott and others has shown that the amount of drift varies according to the season. The large numbers of invertebrates in the drift during spring and summer account for variations in the density of the benthos in a stream. There may be several reasons for drifting. For instance, in some species of mayfly, stonefly, and blackfly the drift rate of the larvae and nymphs increases before pupation or before the adult emerge. This is probably due to changes in behaviour which cause them to be dislodged by current.

The diurnal variation in drift, which increases greatly during the hours of darkness, is probably due to behavioural changes rendering the animals more capable of becoming dislodged. Such movement during daylight might mean that they are more conspicuous to predators. Even bright moonlight can depress drift rate.

Baetis rhodani, *Ecdyonurus forcipula* (Figs. 9.3 (a) and 9.2 (d,e)), and other nymphs show a strong negative phototaxis (movement away from light) as well as a strong thigmotaxis (movement in response to touch), their activity reaching a maximum just after sunset, when negative phototaxis no longer operates. They then tend to move to the top of the stones where they can forage in fresh areas for food.

Although light is regarded as the most important factor governing drift rate, temperature and pH may also have their effects.

Experiments carried out by Townsend and Hildrew [9.6] showed that 2.6 per cent of benthic invertebrates moved from one place to another by entering the drift and some showed an even greater drift rate. They found that *Plectrocnemia conspersa* (Fig. 9.9), for instance, one of the net-spinning caddises, had a drift rate of 14 per cent. This is probably due to the fact that *P. conspersa* is a carnivore and moves to areas where there is a high density of benthic prey. The small bivalve mollusc, *Pisidium* sp., which feeds on detrital matter, enters the drift very infrequently, probably due to its burrowing habits.

Other biotic reasons for drift can be the presence of competitors or feeding conditions becoming difficult in high density populations. The latter could be a reason for drift in *P. conspersa* which constructs a large net, for which space is required, and which responds aggressively to intruders by biting, causing the loser to enter the drift. Incidentally, the shape of net constructed varies according to the rate of flow and the depth (Fig. 9.9 (a,b)).

There have been attempts to determine exactly how far individuals move when in the drift but, although most results are inconclusive and not very accurate, it would seem that many do not drift more than 2 m at any one time.

A shift of position every few days, or more frequently, does take place in most individuals entering the drift. These animals may gain an adaptive advantage in so doing by leaving an unfavourable area for places which offer better conditions, fewer competitors, and more food.

In broad terms one could consider drift to be a form of migration and certainly one that benefits both the individual and the distribution patterns of the benthic inhabitants of a

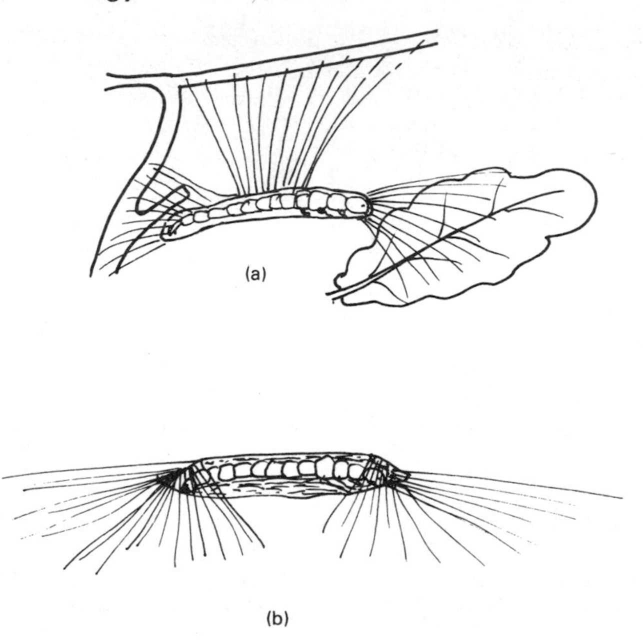

Fig. 9.9 Caddis, *Plectrocnemia conspersa.* The larva rests within a silken tube extended at each end into two asymmetric catching surfaces. The net varies according to flow rate and depth of water. Net constructed (a) in shallow, slow-flowing water (b) in deep, slow-flowing water (both in lateral view)

stream. Migration on a larger scale is, of course, possible for those whose imagos are winged. For crustaceans, molluscs, and the others which are not capable of flight at any stage in their life history, the picture is different. *Gammarus* sp., for instance, shows a high drift rate, tending to move upstream through and over the substratum. The plankton population, on the other hand, is incapable of movement against the current and hence there are higher densities in the lower reaches of a stream, much of this drifting life being swept eventually into the sea.

Fieldwork

Sampling different microhabitats of a stream requires different methods to be employed. It also presents difficulties in standardizing the size of the sample. For sampling the faunal inhabitants of a stream a fine-meshed net is required. Stones are lifted carefully whilst the net is held downstream so that any organisms which are dislodged are swept into the net. It is more difficult to obtain quantitative samples for comparison of numbers and weights of animals present in different waters or at different times of the year. Collections can be made up and across a stream over a certain length of time by lifting stones in the manner just described, and washing off any animals clinging to a stone before discarding it. Although in no way an accurate method, useful results can be obtained if a whole class carries out the sampling to produce mean figures for the different species collected.

1 Collect quantitative samples of organisms in a stream, noting the abundance of the various species. Using the section in the Bibliography 'Ecological Methods', compare your results with other localities and in different seasons of the year. How do your results correlate with the physical and chemical characteristics of the stream?

2 The caseless caddis larvae *Hydropsyche siltalai* and *Plectrocnemia conspersa* are both common stream dwellers occupying different niches. *H. siltalai* spins a rigid net in riffled water to catch detritus while *P. conspersa* is a carnivore spinning a flimsy, irregular net in pools between riffles in order to catch its prey. Using suitable apparatus for the measurement of current (see Chapter 7, Appendices, p. 61), select sites at which both species have made nets and compare the flow rates at a number of these sites. Use references in the Bibliography listed under 'Ecological methods' to interpret your results. Compare the spatial distribution of the two species relative to current flow. The Freshwater Biological Association's publication No. 25 'Caseless Caddis Larvae of the British Isles, by A. G. Hildrew and J. M. Edington will assist in the identification of the larvae.

References

9.1 Elliott, J.M. (1967) 'Invertebrate drift in a Dartmoor stream', *Arch. Hydrobiol,* **63**, 202–37

9.2 Elliott, J.M. (1971) *Some methods for the Statistical Analysis of Samples of Benthic Invertebrates.* Freshwater Biol. Assoc., Scientific publication No. 25

9.3 Ladle, M., Baker, J.H., Casey, H. and Farr, I.S. (1977) 'Preliminary results from a recirculating experimental system: observations of interaction between chalk stream water and inorganic sediment', in *Interaction between Sediments and Fresh Water* (ed. H.L. Golterman), 252–57 The Hague

9.4 Ladle, M., Welton, J.S. and Bass, J.A.B. (1980) 'Invertebrate colonization of the gravel substratum of an experimental recirculating channel', *Holarct. Ecol.* **3,** 116–23

9.5 Roos, T. (1957) 'Studies on upstream migration in adult stream-dwelling insects', *Rep. Inst. Freshwat. Res. Drottingholm,* **38,** 167–93

9.6 Townsend, C.R. and Hildrew, A.G. (1976) 'Field experiments on the drifting colonization and continuous redistribution of stream benthos', *J. Anim. Ecol.,* **45,** 759–72

10 Spatial distribution in stream invertebrates and fish

The phenomenon of invertebrate drift, as we saw in the last chapter, is of great importance when studying the distribution of individuals and populations to the various habitats which supply their needs. In moving from one place to another, however, there are other factors which control distribution. Fish pose particular problems as their movement involves special modifications of behaviour and structure.

The effects of current

For the most part, stream invertebrates rely on the current to bring them their food and to supply them with oxygen. Experiments have shown that some species are only able to perform their life processes within a narrow range of current speed, while others can tolerate a much wider range. But it is difficult to make accurate measurements in the field of the speed of flow exactly at the point where particular species are to be found. As long ago as 1959, an experiment was carried out in which current speed was recorded just above the bottom. It was found that certain species occurred over a wide range of current. The graph in Fig. 10.1. shows some of the results. The nymphs of mayflies belonging to the genus *Baetis* (Fig. 9.3 (a)) were found in greatest numbers at 40 cm s^{-1}, while the more delicate species belonging to the genus *Ephemerella* (Fig. 10.2 (a)) were found in approximately the same

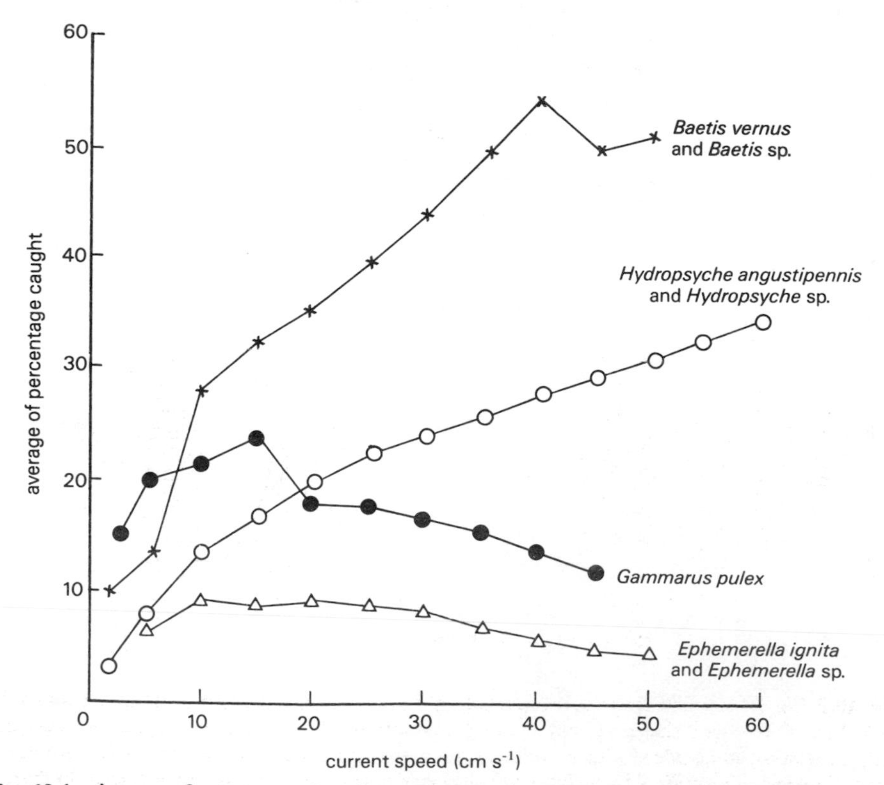

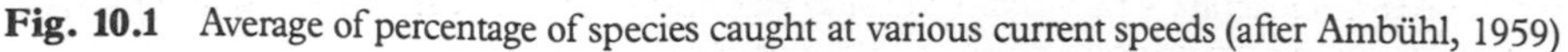

Fig. 10.1 Average of percentage of species caught at various current speeds (after Ambühl, 1959)

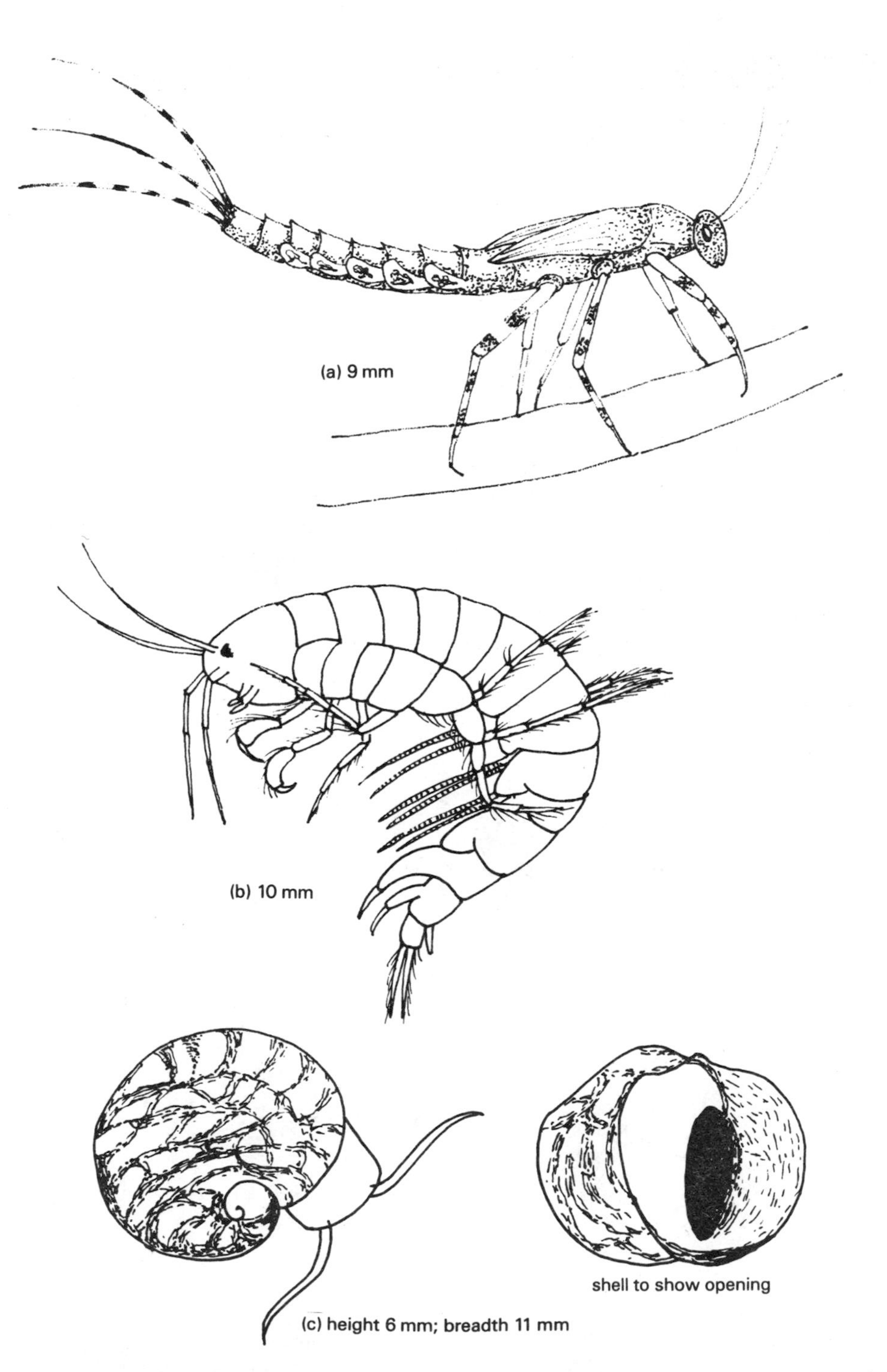

Fig. 10.2 (a) *Ephemerella ignita.* A delicate mayfly found in sheltered waters or where vegetation is thick. In swift reaches it shelters among stones. (b) Freshwater shrimp, *Gammarus pulex,* swims quite well, progressing on its side seeking shelter beneath stones. Prefers well-aerated waters. (c) Freshwater nerite, *Theodoxus fluviatilis.* The shell has purplish markings. The snail has long tentacles. Prefers hard-water rivers and streams with gravel or pebble bottom

numbers from 5 to 50 cm s^{-1}. *Gammarus pulex* (Fig. 10.2 (b)), the freshwater shrimp, was most plentiful at 15 cm s^{-1} while the caddises, *Hydropsyche* sp. (Fig. 10.3), living within their stone shelters, firmly glued to the substratum, reached their peak numbers at 60 cm s^{-1}. Although these figures are useful comparisons they are open to criticism. For one thing the current speed was recorded above the bottom and not actually where the

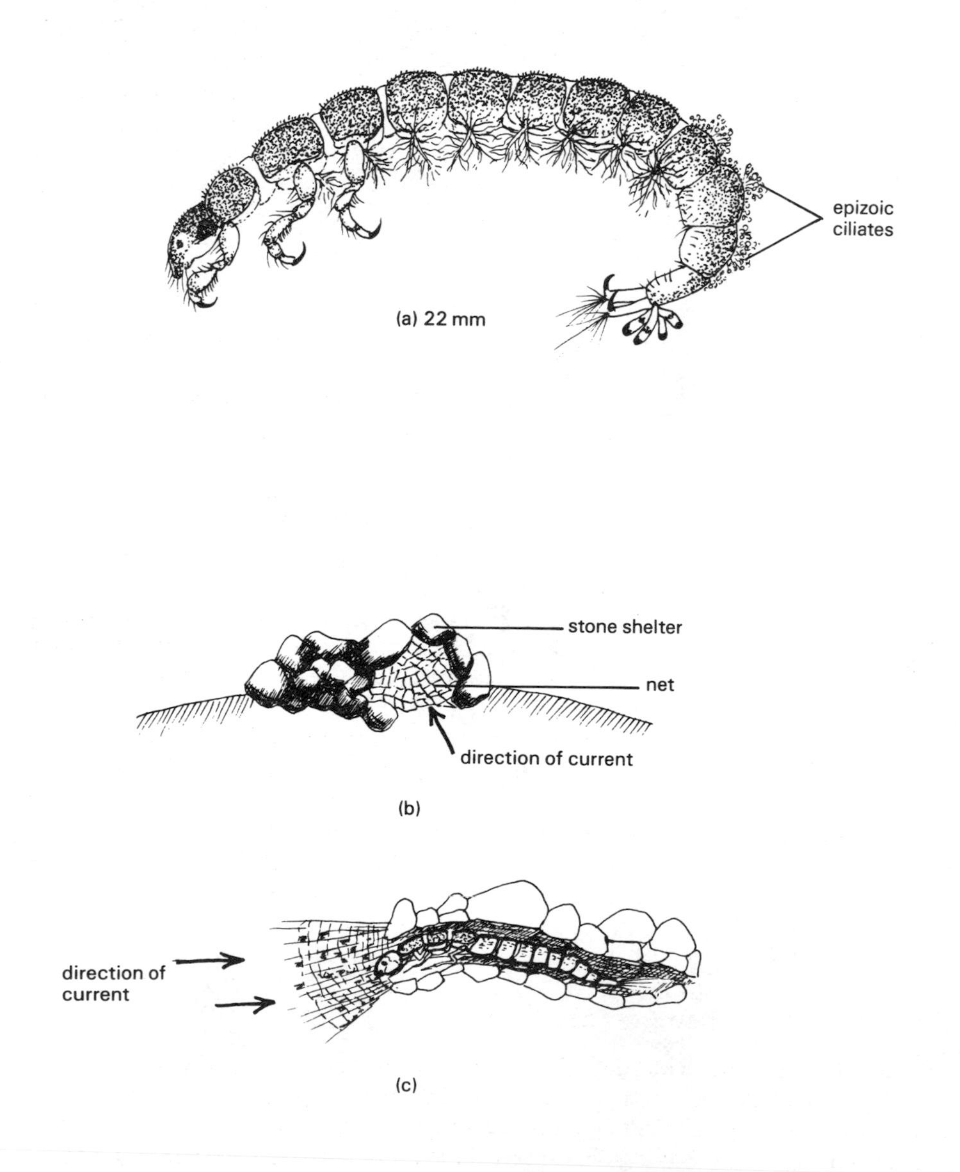

Fig. 10.3 (a) *Hydropsyche angustipennis*. A caseless caddis larva which builds a shelter of stones attached to a larger stone (b). Note the colony of ciliates attached to the last segments. (c) Lying within the shelter, the larva spins a net across the upstream entrance; the silken strands of the net, attached to stones and weed, trap particles of detritus upon which the caddis feeds. Epizoic organisms, such as ciliates and rotifers, benefit from this food current. With each instar the mesh size of the net increases. Nets are usually orientated at right angles to the current, but may be set obliquely in fast flows

organisms were found. Also, except for *G. pulex*, the curves represent a mixture of species. The figures for each curve were calculated as a percentage of the catch of that group, the totals varying for each. This illustrates how faulty methods of sampling can give rise to a misleading interpretation of results.

Where records are made, the relationship between the occurrence of species and their speeds must also take into account the availability of food and the nature of the substratum. For instance, the molluscs *Ancylastrum fluviatile* (Fig. 9.3 (b,c)) and *Theodoxus fluviatilis* (Fig. 10.2 (c)) are both dependent upon the nature of the substratum for a foothold and may, therefore, turn out to be more numerous in swift currents untenable by other species.

It is also interesting that in *A. fluviatile* the shells of specimens found in swift-moving water are consistently taller than specimens obtained from slow-moving water. This phenomenon remains unexplained, and is puzzling since the shells of molluscs living in fast-flowing water are usually streamlined, offering as little resistance as possible to the current.

Hydropsyche angustipennis (Fig. 10.3), a caddis commonly found in flowing water which weights its shell with small stones, feeds by constructing a net to trap passing particles of food. It must, therefore, make a compromise between its ability to hold on and to obtain the best food supply.

Observations have been made of two other caddis species, *Agapetus fuscipes* and *Silo nigricornis* (Fig. 9.5 (b), p. 78), in a spring-fed, uniform and constant stream in Italy where the current speed varied between 7.6 and 38 cm s^{-1}. It was found that although the two species often occurred together over the full range studied, *A. fuscipes* increased in numbers as the current speed increased, while *S. nigricornis* decreased. Furthermore, this slight difference in preference was accentuated in the stationary pupae which positioned themselves respectively in the two current zones.

Species of the larvae of the blackfly, *Simulium* (Fig. 9.4 (b)), have also been found by a number of workers to show different current requirements and also different preferences for positioning on stones and so forth in the substratum, some selecting the top and some the sides of stone, but always facing upstream.

From the large number of field records available, it is clear that it is impossible to state the exact current speed requirements of an individual species. While a current is required by most stream invertebrates, some cannot withstand being actually in it and while current speed is an important factor in the lives of all stream organisms, and controls not only their distribution but also their abundance, it is a complex factor, operating in concert with other factors in their physical surroundings.

The effects of temperature

A rise or fall in temperature immediately alters the amount of oxygen in the water. But other factors have to be taken into consideration, one of the most important being altitude. The flatworm, *Crenobia alpina* (Fig. 10.4), is typical of upland streams and is a relic of the last Ice Age. As long ago as 1912 it was found that, in streams in Western Germany, *C. alpina* occurred only where the temperature did not rise above 16°C. In the lower reaches another species of flatworm, *Polycelis felina* (Fig. 9.2 (a,c)), replaced *C. alpina* until the temperature rose to 16°C and was thereafter replaced by a third species, *Dugesia gonacephala* which is not, however, native to Britain. These observations were regarded as finite for many years until further work showed that *C. alpina* could certainly tolerate temperatures as high as 22°C. Nevertheless it tends to avoid high temperatures by moving upstream during summer to

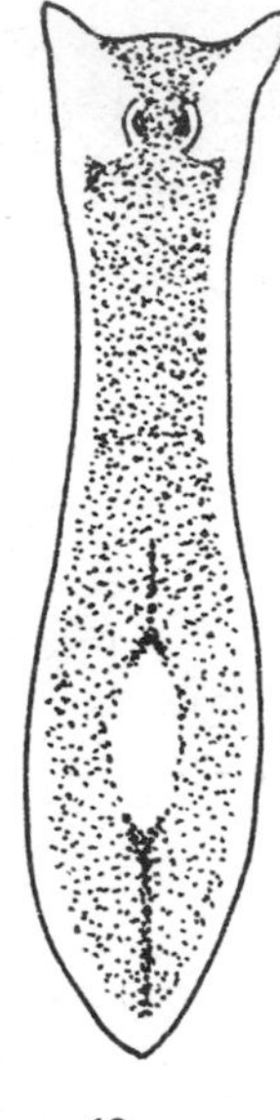

10 cm

Fig. 10.4 *Crenobia alpina.* A flatworm with two eyes usually associated with cool, upland waters (but see p. 89)

cooler regions. In the light of more recent work it is evident that factors other than temperature, such as current flow and pollution, are important in the distribution of flatworms.

The whole question of temperature and its effects on distribution is closely associated with current speed, exposure, and oxygen supply. What is clear is that, taking the seasons as a whole, temperature certainly controls the rate at which the life cycles of many invertebrates progress. In general, low winter temperatures slow down growth rates and the longer the winter lasts, the less time remains for further development to occur.

It is hard to generalize when defining the effects of temperature on distribution since temperature is so closely bound up with other factors, which also include the geographical range of the animal concerned. For instance the river limpet, *Ancylastrum fluviatile*, is almost ubiquitous in hard substrate waters in Europe, whereas in the Carpathians and Russia it is reported to occur only where the water temperature does not exceed 10°C. Why this is so warrants further investigation.

Very cold water, derived from melting ice, and the warm water of hot springs promote enormous changes in the faunal communities. The fauna living around the warm effluents from water cooling towers of power stations also support interesting populations (p. 144).

To breed successfully, and so to produce sufficient numbers of surviving offspring, is a necessary factor in the establishment of a population. But many organisms have an optimum temperature range above or below which they are unable to breed. This may be important, especially for those species which produce more than one generation a year. The blackfly, *Simulium*, lays its eggs on stones in swift-flowing streams. A first batch of eggs may give rise to numerous larvae early in April when the water is relatively cool. Their chief predators are plecopteran and ephemerid nymphs which usually make their appearance a month or so later, ready to attack the second batch of *Simulium* larvae (Fig. 9.4 (b)).

Although some species have a wide range of temperature tolerance, they probably do not occupy the whole of that range. This may be due to the presence of other species existing in competition, at the upper and lower limits of their range, and which have a different temperature optimum. At this temperature, the second species may breed faster or be more active and better able to find and utilize food.

Most of the work done on temperature and its effects on the distribution of organisms concerns their rate of development and therefore their ability to increase and to colonize certain areas in which one species may be in competition with others. For organisms with a complicated life history, such as insects, it is necessary to complete a stage in such a cycle within a certain period of the year. In addition, it may be necessary to produce large numbers of offspring in order to compete successfully with a rival species or in order to maintain a population of sufficient size in face of predation.

Distribution in different substrata

Regarding the choices open to invertebrates, there are many reasons for the selection of one substratum rather than another. It is all too easy to jump to conclusions which may not reflect the whole picture in a particular situation, nor will the reactions of a species to a single substratum necessarily be the same in all streams. Sessile animals, such as bryozoans and sponges, select a solid surface, perhaps a large stone or log, rather than shifting gravel. *Ancylastrum fluviatile* also prefers a stable surface. The mayfly, *Ephemera danica* (Fig. 10.5 (a)), on the other hand, prefers a sandy substratum with a particle size of 0.05–3 mm and it becomes noticeably scarcer as the particle size increases. This preference is associated with the burrowing habits of the nymphs and their ability to keep their burrows open.

In general, the more complex the substratum, that is the greater the range of variation from large stone to sand and silt, the more diverse will be the invertebrate fauna. Sand is relatively poor in numbers of species, silty sand being richer, and muddy sand supporting larger populations but being poor in numbers of different species.

In Britain two species of stonefly, *Dinocras cephalotes* and *Perla bipunctata* (Fig. 10,5 (b, c)), are often found in the same stream but in different habitats. *D. cephalotes* usually occurs in the areas of stable stone bottom where its dark colouring may offer protection, while *P. bipunctata* tends to live in the areas of shifting stone bottom where its body colour merges with the stones.

Water moss, filamentous algae, and other plants growing on stones, waterfalls, and in the stream bed, greatly affect the number of invertebrates colonizing these areas. The plants offer both shelter and a foothold, as well as possibly increasing the oxygen content in the area immediately around them. By far the most numerous animals usually to be found living among these plants are several species of chironomid larvae which construct tunnels of silt glued together.

Vertical distribution in the substratum

In chalk streams, usually containing large quantities of fine sediments, it is doubtful if many invertebrates live much deeper than 100 mm, except where upwelling or throughflow of ground water occurs [10.6]. On the other hand, Williams and Hynes [10.7] found that a large number of individuals of many species occurs quite deep in the beds of streams with coarse sediments, living in the interstices between the stones. These animals are termed collectively **hyporheos**. Work was done in the Speed River in Canada where monthly samples were taken with a core sampler at 10 cm intervals down to a depth of 70 cm. Few

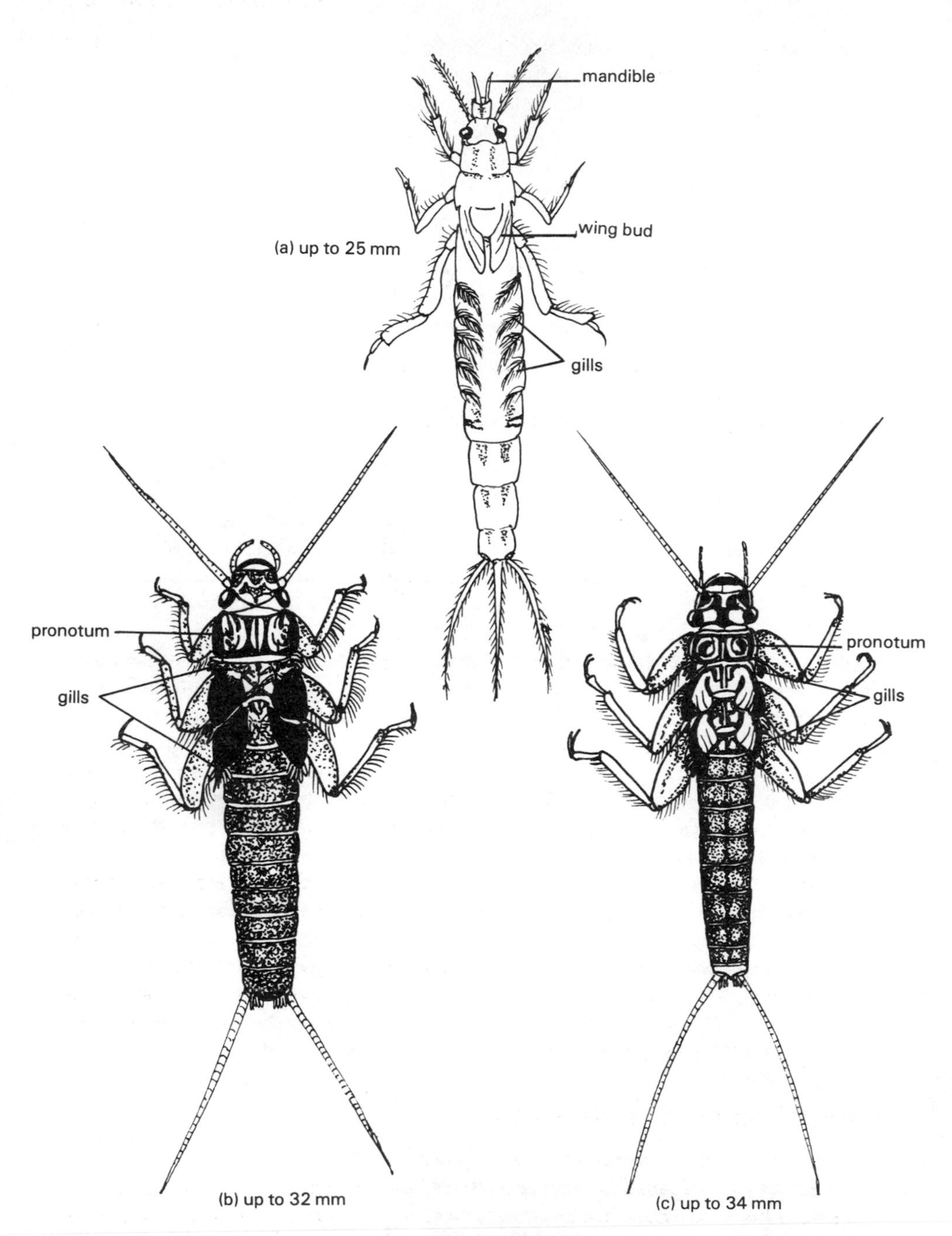

Fig. 10.5 (a) Mayfly nymph (Ephemeroptera), *Ephemera danica.* The gills are held flexed over the abdomen. Burrows into sandy beds of streams by means of the strong jaws and legs, feeding on diatoms and other algae. Very sensitive to light. (b) and (c) Two stonefly nymphs (Plecoptera). (b) *Dinocras cephalotes* is dark brown with yellow patterns on the thorax. Pronotum twice as wide as long. Last abdominal tergum brown. Inhabits swift streams with a stable substratum. (c) *Perla bipunctata* is black with bold yellow patterning. Last abdominal tergum yellow. Pronotum less than twice as wide as long. Inhabits swift streams with unstable substrata. Both species distinguishable from other plecopterans by the presence of gills on the thorax

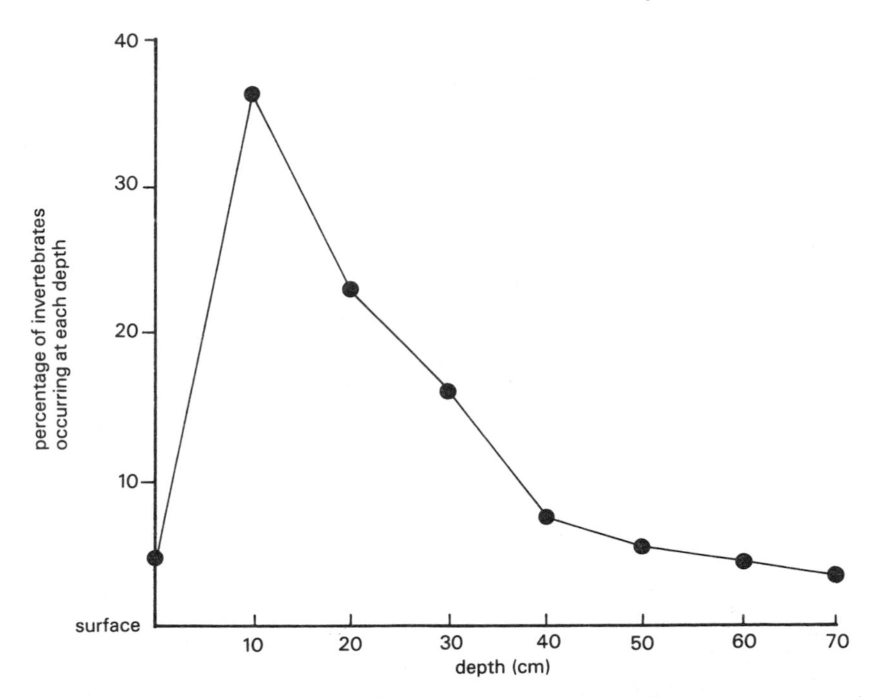

Fig. 10.6 Mean percentage depth distribution within the stream bed for all invertebrate taxa in the Speed River, Canada (redrawn from Williams, D.D. and Hynes, H.B.N. (1974) *Freshwat. Biol.*, **4**, 233–56)

invertebrates were found below this depth. Figure 10.6 shows the mean percentage depth distribution for all taxa combined. The maximum density occurred just below the surface and decreased with increasing depth.

Some hyporheic organisms, such as species of mites and micro-crustaceans, live permanently as members of the hyporheos, while others seek out this community for only part of their life history. Williams and Hynes found that the larvae of one genus of chironomid (*Cladotanytarsus*) overwintered in the hyporheos, possibly seeking an optimal temperature for growth, and migrated upwards when the surface water began to warm up in the spring. It is also probable that some members of the invertebrate fauna may temporarily move deeper into the substratum during flood conditions.

Recolonization of substrata

An area of a stream bed can become denuded of organisms due to catastrophic drift but will quickly become recolonized. The time taken to regain normal density is often only two or three weeks. This is evidence of the large-scale movement of organisms that is occurring all the time.

Williams and Hynes [10.7] experimented in a slow-flowing Canadian stream in order to determine the degree of importance of the various types of movement in recolonization. They assumed that there were four chief sources of colonists: invertebrate drift, upstream movement of animals in contact with the bottom, an upward vertical movement from deep down in the sediments, and colonization by oviposition from aerial sources. Four types of traps were constructed, designed to isolate each of the four categories of colonists (Fig. 10.7).

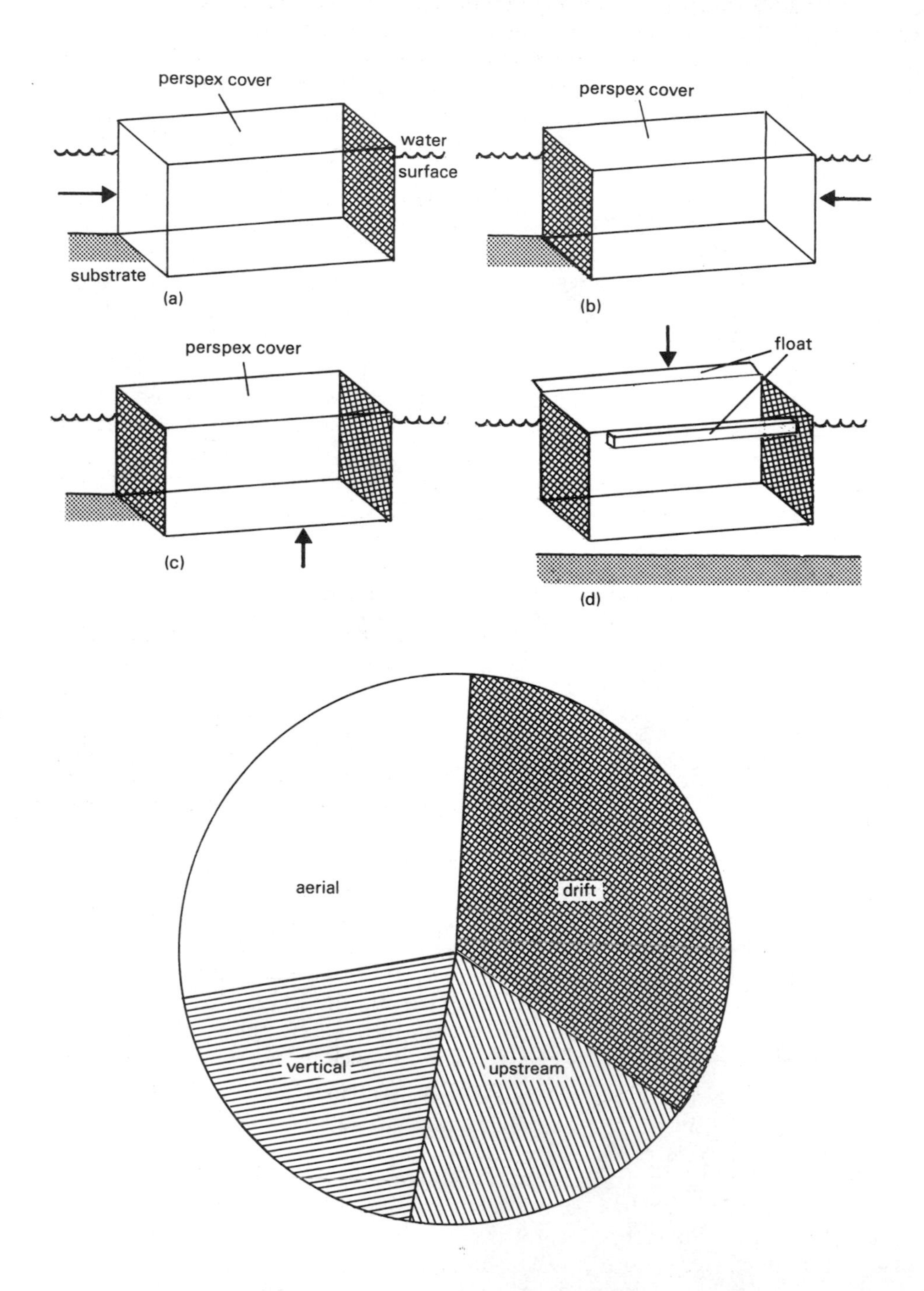

Fig. 10.7 Pie chart showing the proportionate colonization of traps by (a) drift (b) benthic upstream migration (c) vertical upward migration (d) aerial sources of traps (a) (b) (c) and (d) above [see text for description]. In all cases, streamflow is from left to right

The traps were wooden boxes 60×30 cm. Trap (a), designed to prevent access except by invertebrate drift, had a perspex cover to prevent aerial oviposition and the upstream end was left open. But an experimental error was introduced by placing the trap on the stream bed so that it also collected animals moving downstream along the stream bed as well as those in the drift. Trap (b) had a mesh screen in the upstream end but was open at the downstream end. In this way it permitted colonization by upstream moving invertebrates only. Trap (c) had screens at both ends but allowed invertebrates to move vertically into the colonization tray from the substratum beneath. Trap (d) floated on the surface with the lid removed to allow access only to aerial ovipositioning insects. Two traps of each kind were placed in the stream in June and left for 28 days. Two controls were constructed which allowed access from all sources. Results showed that at the end of the experimental period 34 000 invertebrates colonized the control trays and these were principally chironomid larvae. Proportional results for the other trays are illustrated by the pie chart in Fig. 10.7. This shows that although all four sources contributed to colonization, by far the most important were those from the drift.

A similar experiment was repeated in a modified form by Townsend and Hildrew [10.5] which eliminated the error in the William Hynes trap (a) by placing the colonization trays on platforms 5 cm above the stream bed. These could only be colonized by drifting organisms. Results revealed that more than 80 per cent of invertebrates migrating from one area to another in all probability did so in the drift. These experiments clearly show that invertebrate drift is of prime significance in the distribution of stream benthos.

Habitat Selection

Every micro-habitat in a stream offers a number of different environmental factors to which successful organisms living in that habitat have had to adapt. The question arises as to whether the spatial pattern of a species correlates with a particular factor or whether such a pattern of distribution is due to several environmental factors. Controlled laboratory experiments, isolating one factor that varies while others are held constant, can show that correlation between distribution and a particular factor is due to behavioural selection of one factor (such as the presence, for instance, of the organism's natural food) or to the nature of the substrate.

Such a possibility has been shown in a number of investigations using the freshwater shrimp, *Gammarus pulex*. In these studies depth and rate of flow did not provide a convincing explanation of the often 'clumped' distribution of this shrimp. On the other hand the distribution *was* significantly related to the nature of the substratum and age of the individuals. While the smaller and presumably younger shrimps were distributed across the whole range of mineral particle size, avoiding only the fine silt, the larger individuals were restricted to areas where the particle size was larger. This is probably because they require spaces between particles of the right size in which to wedge themselves, thus protecting themselves from the current and predators. Clumping was found to be due to the larger gammarids moving about more slowly in substrates of the preferred size and thus collecting in appropriate areas.

Detrital matter on which *Gammarus* feeds also causes large numbers of individuals to accumulate in one area, showing that either particle size or the presence of food are both factors influencing their distribution.

Chemical factors and distribution

In flowing water, oxygen is seldom a problem, except in polluted regions, because it is usually well aerated. The existence in some slow-flowing stretches of a stream of a deep layer of rotting leaves or other vegetation may create a local oxygen lack. Likewise in the backwaters of a stream or river there may be dense stands of emergent plants creating a stagnant area similar to marshland associated with lakes. Here oxygen levels will be low and the invertebrate inhabitants will be those associated with ponds.

It is doubtful, despite observations made on many invertebrates both in this country and abroad, whether acidity operates as a controlling factor in their distribution. Nevertheless, acid waters support a rich invertebrate fauna, while chalk streams (which are hard waters) have their selection of invertebrates especially those, such as molluscs, which secrete shells composed of calcium carbonate, and freshwater amphipods and isopods.

Osmotic control in invertebrates

The body fluids of most freshwater invertebrates have a lower osmotic pressure than that of the surrounding water and therefore there is a continual passage of water into the body through any permeable surface.An impermeable body covering of chitin or cuticle reduces the areas of those surfaces in many species, such as insects and crustaceans. Because of the problem of water intake, these animals have evolved efficient means of eliminating excess water by developing appropriate urinary systems producing large quantities of dilute urine which, by excretion, rids the body of excess water.

The situation is different, however, in waters which have a greater salinity than the body fluids of the invertebrates living there. Under such circumstances the reverse process may take place — the loss of body fluids to the surrounding water. Some animals are capable of withstanding a wide range of salinity. They are the **euryhaline** species living in estuaries between fresh and salt water. A few, like the mitten crab, *Eriochier sinensis* (Fig. 10.8) do not produce a dilute urine and cannot therefore cope with much fresh water. Incidentally, this often quoted species, an import from China, is now on the British list. It does not often turn up but one specimen was found in the Thames in 1935, and in 1976 three crabs were found in the cooling water screen of a power station on the Thames estuary. They probably arrived from China as larvae in the ballast tanks of commercial ships and the discharge of warm spent cooling water may have been at the right temperature for their further development to adult crabs.

Some species of *Gammarus*, the freshwater shrimp, show an interesting series of tolerance to saline conditions. *G. pulex* occurs in rivers down to the point where the fresh water comes into contact with sea water. Another species, *G. zaddachi zaddachi*, is also found in fresh water, although it never breeds there, and extends down the estuary to a point where the salinity at mean high tide is 10–15 per cent. The subspecies, *G. zaddachi salinus*, takes over in the middle reaches of the estuary and can tolerate a salinity a little less than that of sea water while *G. locusta* is a common shore species, tolerating dilution up to 17 per cent.

The freshwater winkle, *Potamopyrgus jenkinsi* (Fig. 9.6), has evolved only recently from a marine environment through our estuaries to populate nearly all inland bodies of water. Problems of osmoregulation must have been resolved in this case by high selection pressure on genotypes of the pioneer species.

Returning to the question of hard and soft waters, the degree of hardness seems to be an important factor in the distribution of certain animals. Reynoldson [10.4], working on planarians collected in different lakes, showed that some species have a distinct distribution pattern according to the concentration of calcium (Fig. 10.9) and this may also be so in flowing waters showing different degrees of hardness.

Hard water, because it contains more calcium and also more of several other ions such as chloride, sulphate, magnesium, and sodium, has a higher osmotic potential than soft water. Invertebrates overcome this problem by absorbing the salts, thereby increasing the

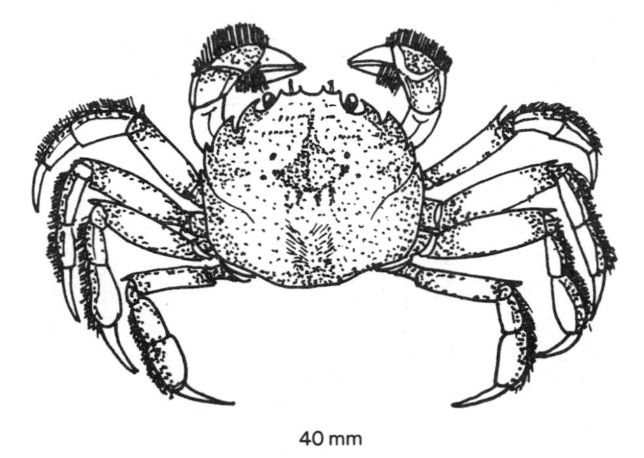

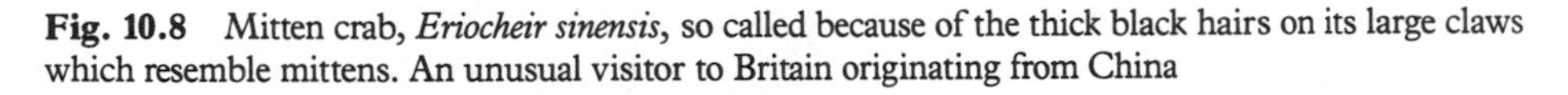

Fig. 10.8 Mitten crab, *Eriocheir sinensis*, so called because of the thick black hairs on its large claws which resemble mittens. An unusual visitor to Britain originating from China

concentration of their body fluids and lessening the tendency to take in water by osmosis. Soft water contains fewer salts and therefore has a lower osmotic potential. This can present problems of osmoregulation, as in the mitten crab already mentioned. Although we know that water hardness does control the distribution of at least some animals, notably molluscs, field work has produced conflicting results and we need to have more carefully collected and assessed records before we shall know exactly how hardness exerts its effects.

Floods and droughts

In rivers and streams, where flooding is a regular seasonal feature, there tends to be a corresponding seasonal change in the numbers and species of invertebrates present. It is also true that such streams have a less abundant and less varied fauna than others.

Sudden spates can cause a rapid rise in the rate of flow, far beyond that against which many of the invertebrate inhabitants can maintain a foothold. This can result in large numbers being washed out of a stream. In August 1968 the total population of *Potamopyrgus jenkinsi* was swept out of a stream in East Devon into the Exe Estuary. Doubtless many other species suffered the same fate during this sudden flood, especially the stonefly and mayfly nymphs normally living on the stony bottom. Those more securely attached, like the freshwater limpet, *Ancylastrum fluviatile*, and the leech, *Erpobdella octoculata*, survived with little depletion in numbers.

Drought is another hazard [10.1] and it can be said that not many species are capable of taking effective steps to cope with this emergency. Of course it is more usual for streams to dry up in summer than at other times of the year and by then many insects passing their immature stages in water may have flown, after mating and egg-laying. So we find that in a number of species it is the egg stage which can best withstand drying up. The caddis, *Agapetus*, on the other hand, has a fairly resistant pupa which, if exposed to drought, seems capable of successful emergence. Some leeches, too, affix their hard cased egg cocoons to stones and can withstand being dried up in this condition. Strangely enough, after a drought new colonists often appear, apparently arriving from nowhere. The blackfly, *Simulium ornatum* (Fig. 9.4 (b)), is one such opportunist which can move into a recently dried-up area, pass through one generation fairly rapidly, and then disappear again. An explanation for this is probably that the imago lays resistant eggs which survive drought.

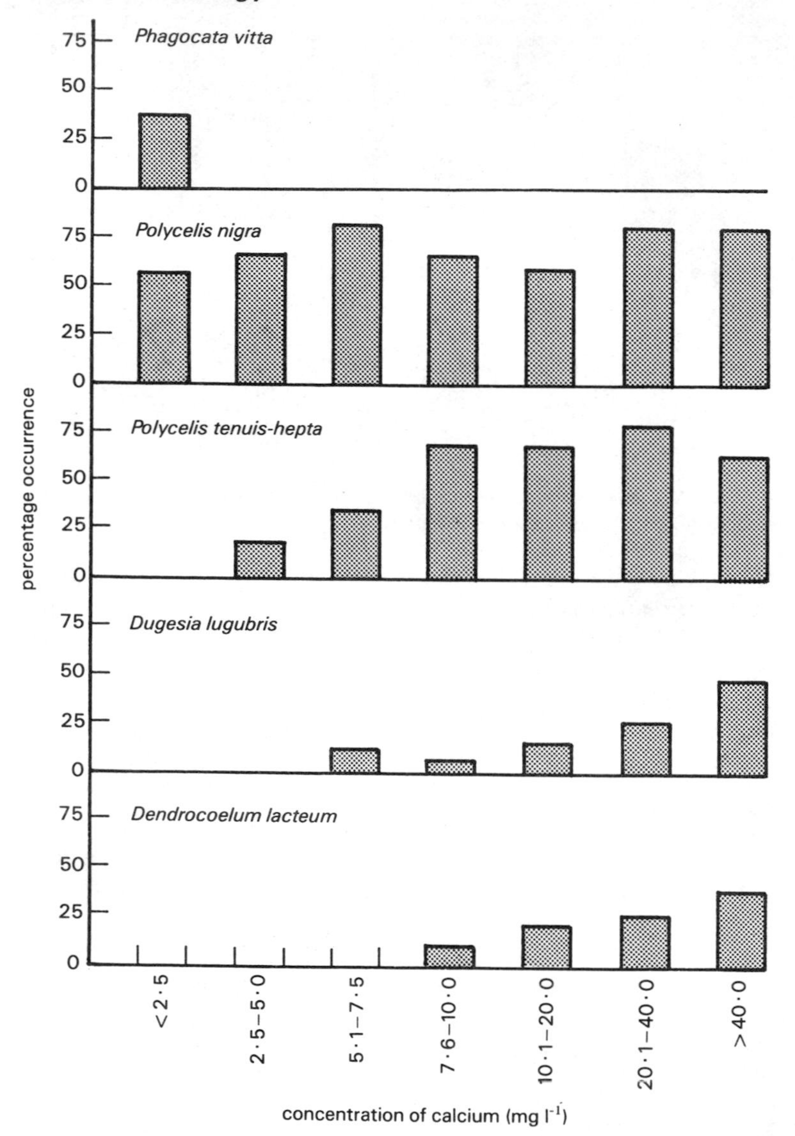

Fig. 10.9 Percentage occurrence of five species of triclad in waters with various concentrations of calcium (after Reynoldson, 1958)

Fish and their adaptation to flowing water

Fish living in running water are adapted in many ways to the physical, chemical, and biotic factors of their environment. The brown trout, *Salmo trutta*, has a body round in cross-section and streamlined offering little resistence to the current (Fig. 10.10 (a)). Although trout are strong swimmers and in cases of danger can dart at great speed, most normally swim in short bursts, resting for long periods on the bottom or taking shelter behind stones where the water is calm. The tail is the main means of propulsion combined with the flexing of the body muscles from side to side (Fig. 10.10 (b)) and steered by the unpaired dorsal fins. Vertical movement is brought about by the paired pectoral and pelvic fins.

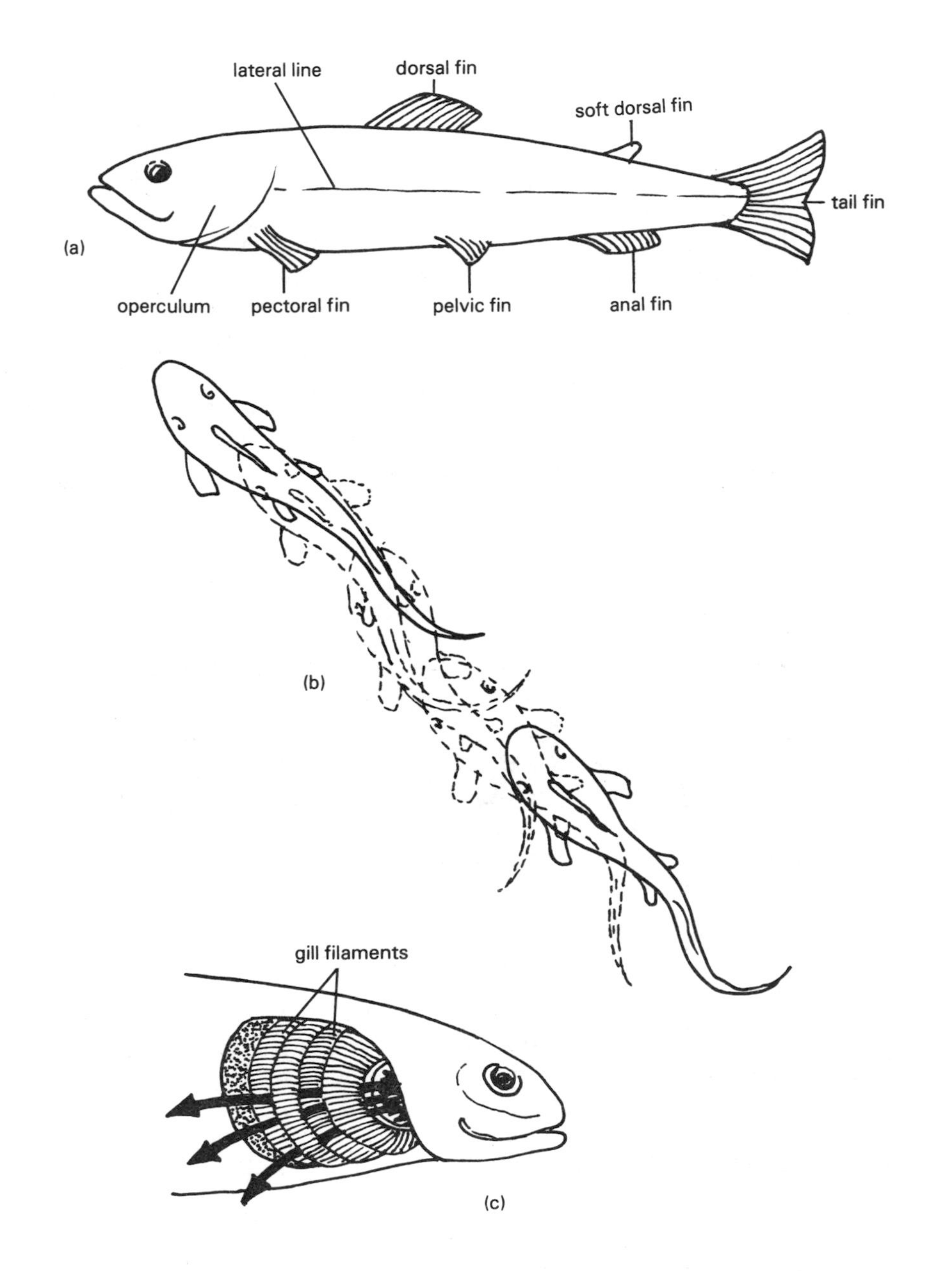

Fig. 10.10 Brown trout, *Salmo trutta*, (a) to show the streamlined body and the arrangement of the fins; (b) movement through the water is achieved by pushing aside the water with the forward movement of the head followed by the body flexing from side to side and finally the tail; (c) head of fish with operculum removed to expose the gill filaments. Arrows show the course of the water as it flows over and between the filaments where gaseous exchange takes place

As with all totally aquatic organisms, fish are dependent on oxygen dissolved in the water for their respiratory functions. By closing the gill covers (operculae), opening the mouth, and at the same time depressing the floor of the mouth, pressure inside is decreased and water rushes in. By closing the mouth and opening the operculae, the water is forced out, passing over the gills which are richly supplied with blood vessels. Here an exchange of gases takes place, oxygen from the water entering the blood and carbon dioxide leaving it by diffusion (Fig. 10.10 (c)). Alternate opening of the mouth and gill covers takes place all the time, both at rest and during periods of activity, for fish require a constant supply of oxygen at all times. The rate of respiration increases if the oxygen content of the water is depleted. Some fish regularly rise to the surface to gulp air.

The stone loach, *Nemacheilus barbatulus* (Fig. 10.11 (a)), and the bullhead, *Cottus gobio* (Fig. 10.11 (c)), are typical bottom-living species, spending all their time close to the stream bed. They are dorso-ventrally flattened with eyes placed on top of their heads, giving clear vision of what is going on above the substratum. In addition, *N. barbatulus* possesses pairs of sensitive barbules beneath the head which probably assist in testing the substratum for food and sensing passage through weeds (Fig. 10.11 (b)). The bullhead is helped in maintaining its position on the bottom by a kind of friction plate formed by the pelvic and pectoral fins being brought together (Fig. 10.11 (d)).

Colour is important for fish as a means of camouflage. Benthic species, such as the bullhead and stone loach, adopt a mottled body colour to match the stone substrata and are extremely difficult to see unless they move.

All fish have a lateral line system (Fig. 10.10 (a)) by means of which they are aware of vibrations in the water. The lateral line system, with its sensitive nerve endings, is used in echo-location and is also a fish's distance receptor in muddy water.

The swim-bladder is an organ found in all bony fish. It is filled with air and lies internally, above the gut and just ventral to the spine. It is particularly important in buoyancy control, enabling trout, pike, loach and many other river fish to adjust their position vertically in a body of water. This adjustment is achieved by inflation or deflation of the swim bladder. In benthic species the bladder, not surprisingly, is greatly reduced, giving them the negative buoyancy they require. It is interesting that when frog tadpoles are found in torrent waters they develop very small lungs, presumably serving to reduce buoyancy.

All river fish are easily fatigued, hence the long periods they spend resting, which we have already mentioned (p. 98). Fish muscle, like mammalian muscle, manufactures lactic acid during a period of activity and oxygen is used up. Lactic acid accumulates much faster in fish than in mammalian muscle and also takes longer to disperse. Hence the lengthy periods of resting required by fish.

Temperature, the oxygen content of the water and the nature of the substratum all interact to influence the distribution of fish in a stream and of course the rate of flow is also an important factor.

Fish expend much energy in fighting the current so it is not surprising that there are differences between species in their ability to do so. Below is a list of the maximum swimming speeds of a number of species. These are closely related to the relative rate of flow in the places they inhabit:

Salmon, *Salmo salar* (adult) 800 cm s^{-1}
Trout, *S. trutta* 400 cm s^{-1}
Chub, *Squalius cephalus* 270 cm s^{-1}
Bream, *Abramis brama* 55–65 cm s^{-1}
Tench, *Tinca tinca* 45–50 cm s^{-1}
Pike, *Esox lucius* 45 cm s^{-1}
Carp, *Cyprinus carpio* 40 cm s^{-1}

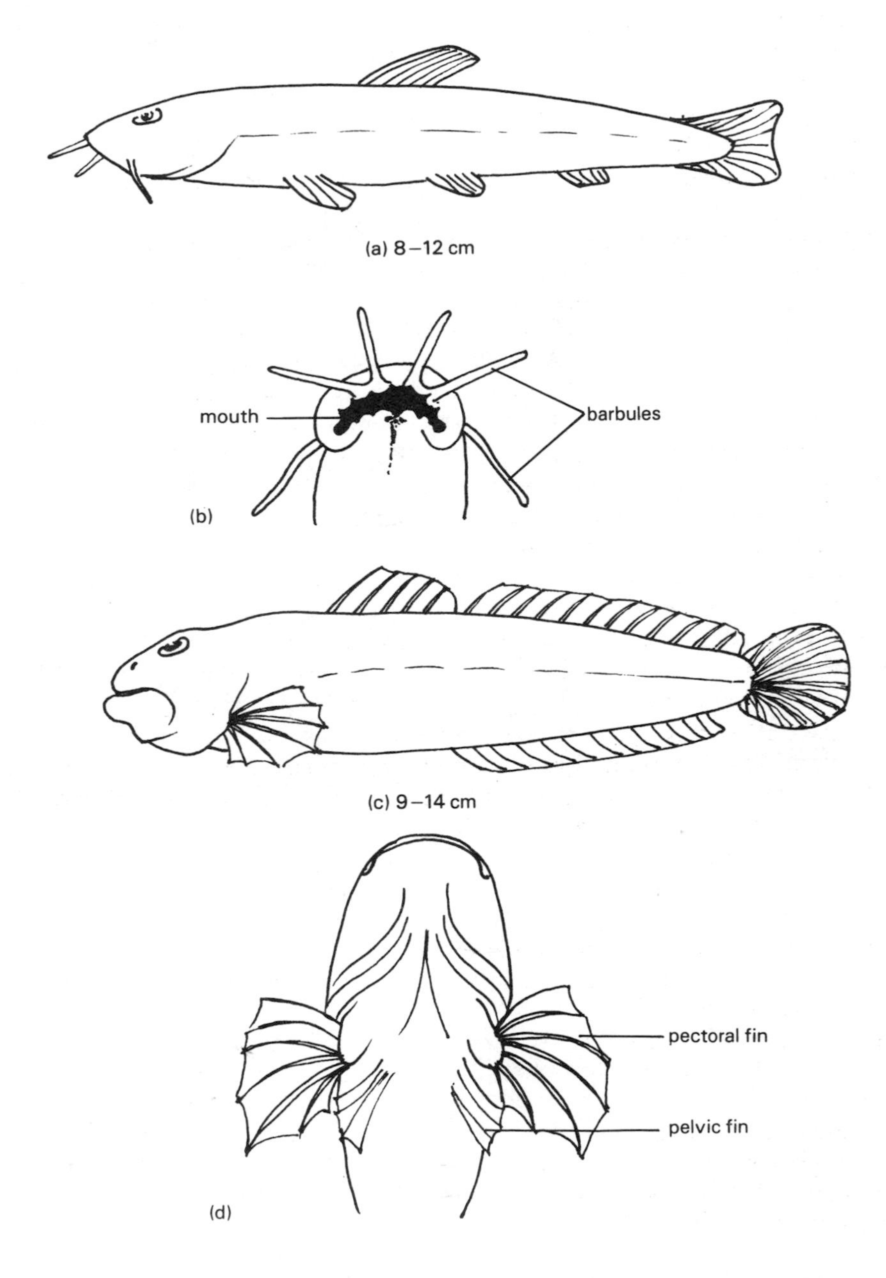

Fig. 10.11 Two benthic freshwater fish. (a) Stone loach, *Nemacheilus barbatulus.* The eyes are on top of the head. (b) Ventral view of head of stone loach, showing ventrally placed mouth surrounded by barbules. (c) Bullhead, *Cottus gobio,* in which the eyes are on top of the head but the mouth is not ventral. (d) Ventral view of head of bullhead, showing friction plate formed by the pelvic and pectoral fins, which helps to anchor the fish beneath stones in a fast current

Osmoregulation in fish

A major difficulty for organisms invading rivers from the sea is the difference in the osmotic conditions prevailing in salt water and fresh. This may be a suitable point at which to give a brief account of what is meant by osmoregulation in fish and what is involved for the species concerned.

Fish usually found only in fresh water have, like the simpler organs of invertebrates, efficient urinary systems which excrete considerable quantities of dilute urine. Due to the amount of energy involved in maintaining a constant internal osmotic pressure, freshwater fish use much oxygen. On the other hand, fish living in the sea lose water to the outside, their body fluids being hypotonic to it. To counteract this, they drink quantities of sea water and excrete a concentrated urine, excess salts being eliminated through special cells in the gills. Some, like the sea lamprey, salmon, and eel, are capable, by adjustment of their osmotic systems, of living in both sea and fresh water at different periods of their lives.

The eggs of freshwater animals often attain complete impermeability to water by a waterproof membrane. This is certainly true of trout and for a number of other species. At later stages the membrane may become slightly permeable, since the intake of water would seem to be a biological necessity for the growing embryos. The eggs of frogs and those of cladocerans are not completely impermeable, since a slow intake of water is needed all the time for the development of the embryo.

Migration

Migrations to or from the sea occur in several species of fish in Britain. As mature fish, salmon return to the river of their origin, making their way upstream to suitable spawning grounds. Here the female constructs a redd (spawning-bed) in patches of gravelly sand and lays her eggs. The male then fertilizes the eggs. Many eggs will be washed away in the current, despite the sticky covering surrounding each one.

Another migrant from sea to river is the sea lamprey, *Petromyzon marinus* (Fig. 10.12). The extraordinary history of the lamprey is less well known than that of the eel (*Anguilla anguilla*) which also migrates from sea to river.

Besides *P. marinus*, two other species of lamprey occur in our rivers: the parasitic river lamprey, *Lampetra fluviatilis*, and the non-parasitic brook lamprey, *L. planeri*. The life cycle of all three species is divided into two distinct phases, the larval form, or ammocoete, and the adult. Ammocoetes are blind, filter-feeding organisms living concealed in river silt. Pickering [10.3] describes the osmoregulatory mechanism in the river lamprey which, for both river and marine species when they are in fresh water, involves copious urine production to balance the large intake of water. After a period of 3–5 years, the ammocoete of *P. marinus* undergoes a metamorphosis into a sexually immature, non-feeding stage (a macrophthalmia) which is very active and capable of osmoregulation in either fresh or sea water. In the sea the lamprey is subject to a continuous osmotic loss of water which is balanced by the ingestion of sea water. The migration back to rivers for spawning involves another osmotic adjustment and feeding ceases, the lamprey's total metabolic demands being met from stores within the body tissues.

The adult *P. marinus*, measuring some 70 cm or more, presents a somewhat bizarre appearance as it struggles upstream, for in doing so it can get turned over sideways to display its bright yellow underparts. This colour is more suited to the marine environment and renders the animal particularly obvious against the dark stones of a river bed. Sea lampreys use a number of rivers around our coasts for spawning, the River Otter in Devon being one. They usually enter the river in late June or July and make their way up to the limit of the tidal

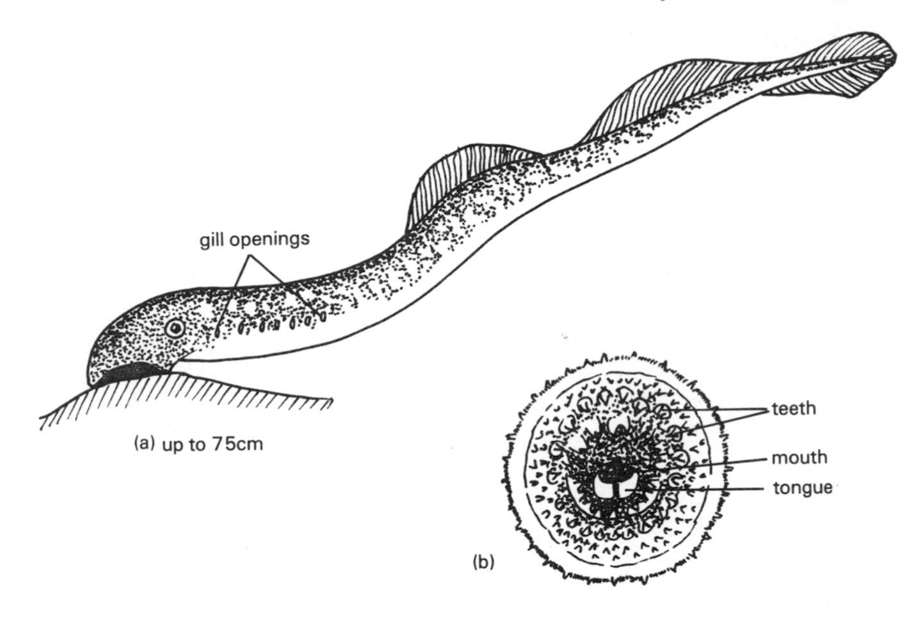

Fig. 10.12 (a) Sea lamprey, *Petromyzon marinus.* In fresh water the sucker is used solely as an anchor, but in the sea it parasitizes a wide variety of marine fishes. (b) Ventral view of mouth, containing up to 125 sharp teeth which, with the toothed tongue, are used for rasping and drilling through the scales of its fish host. Glands in the mouth exude anti-coagulant which keeps the host's blood flowing

waters some 2½ km upstream. Here they construct a large 'nest' in the bed of the river using the powerful mouth sucker, designed to affix them to their marine hosts, to move stones downstream in a wide semi-circle about 80 cm across. Clearing the 'nest', they can lift quite large stones to the perimeter, leaving a bottom of small pebbles. The female lays her eggs amongst the pebbles which are fertilized by the male, always in attendance. The eggs are sticky and as they are washed downstream towards the wall of the 'nest' they adhere to the stones. Many will soon be devoured by eels moving among the stones but some will survive to ammocoete stage. The mature lampreys, once they have spawned, fall back spent, to be washed downstream to die. The objection taken to them by river bailiffs and others protecting fishing is ill-founded since *P. marinus* does not feed whilst on its breeding journey upriver, during which it becomes thinner and shorter.

Fieldwork

1 Many streams have large populations of the freshwater shrimp, *Gammarus pulex*, some being smaller young specimens and some larger mature individuals. Select two areas of the stream bed, one which is sandy and one composed of larger particles. Collect specimens of *Gammarus* from each. Do you notice any difference in the average size of individuals from each site? Record the rate of flow on the stream bed at each site using one of the methods described in the appendices to Chapter 7. Can you relate any difference in size of *Gammarus* to the velocity of the water?

2 Continuing the study of the distribution of *Gammarus* in the stream, try to discover if there are other factors affecting distribution. Place a quadrat, 25 cm square, in an area of

the stream which is clear of debris and count the number of individuals within the quadrat. Repeat the process in an area where leaf litter, the food of these detritivores, has accumulated, holding a net just downstream of the quadrat while disturbing the substrate to dislodge any hidden specimens. Are the numbers greater and, if so, is there any connection between your results in both these investigations? You may have to repeat these experiments in five or more areas.

3 The experiments carried out by Williams and Hynes, described in this chapter, did not completely isolate invertebrate drift as a colonist source. Investigate this further using two sets of trays measuring, say, 37×17.5×2.5 cm. Place one set actually on the bed of the stream, thereby trapping those invertebrates entering from the drift and those walking across the bottom. Place the other set of trays 5 cm above the bottom, so trapping only drift organisms. Every three days select at random four trays from the bottom set and four from the suspended set. Count the number of invertebrates in each set. After twelve days (four counts) record results for each set of trays as a graph, and compare the results. The larger the number of trays used, the greater accuracy you will achieve by this method of sampling. What do you conclude?

Invertebrate drift varies considerably in amount, both seasonally and diurnally. These variations could be investigated using the same apparatus. During the hours of daylight many invertebrates rest immobile under stones or within the substratum, emerging only at night to forage. Other species show a higher drift rate during the day or during certain periods of their life histories (e.g. the larvae of the blackfly, *Simulium*). Such behavioural patterns could also be investigated using sets of trays in the same manner.

References

10.1 Ladle, M. & Bass, J.A.B. (1981). The ecology of a small chalk stream and its responses to drying during drought conditions. *Arch. Hydrobiol.* **90**, 448-66

10.2 Mills, C.A. & Mann, R.H.K. (1983). The bullhead *Cottus gobio*, a versatile and successful fish. *F.B.A. 51st An. Rep.* 76-88

10.3 Pickering, A.D. (1978) Physiological aspects of the life cycle of the river lamprey, *Lampetra fluviatilis* L. *F.B.A. 46th An. Rep.* 40-46

10.4 Reynoldson, T.B. (1958) 'Triclads and lake typology in northern Britain — qualitative aspects'. *Verh. int. Ver. Limnol* **13**, 320-30

10.5 Townsend, C.R. & Hildrew, A.G. (1976) Field experiments on drifting, colonization and continuous redistribution of stream benthos. *J. Anim. Ecol.,* **45**, 759-72

10.6 Welton, J.S., Ladle, M., Bass, J.A B. & Chapman, K. (1981) Invertebrate sampling in the substratum of an experimental recirculating stream. *Int. Revue ges. Hydrobiol. Hydrogr.* **66**. 407-14

10.7 Williams, D.D. & Hynes, H.B.N. (1976) The recolonization mechanism of stream benthos. *Oikos*, **27**, 265-72

11 Problems of gas exchange, locomotion, feeding and reproduction

Obtaining oxygen and getting rid of carbon dioxide as a waste product is one of the greatest problems faced by freshwater animals. It involves a wide variety of modifications, not only of structure but also of behaviour. The close connection between the physiological and chemical conditions prevailing in a body of fresh water brings about a number of responses to these conditions.

Creatures living in fresh water for part or all of their lives can be divided into two main categories. **Partially aquatic** animals are those which live beneath or on the surface of the water and which are dependent directly upon atmospheric oxygen for respiration. **Totally aquatic** animals, on the other hand, are those which live beneath the surface and rely on oxygen dissolved in the water. We shall include in this chapter a third much smaller group, the **anaerobes** which can exist in the anoxic conditions associated with sediments denuded of oxygen or in certain regions where decay or organic debris has caused an almost total oxygen lack.

As we have seen (p. 8), the oxygen content of a body of fresh water is inversely proportional to the temperature. The concentration of oxygen also fluctuates according to depth, to the area of the surface, to the air pressure during the day and night, and from one period of the year to another.

Surface dwellers

If a steel sewing needle is carefully lowered onto the surface of water in a beaker, it will float, supported by the surface film. A heavier object will break through the film and sink. A few small animals, mostly insects, are so light that they can exploit the surface tension and walk or skate about without penetrating it. Not only does their lightness enable them to do this but their bodies, and especially their tarsi, are covered with **hydrofuge** or water-repelling hairs.

A pond skater, *Gerris* sp., (Fig. 11.1 (a)) placed on the surface of water in an aquarium is unable to sink. The lightness of its body, and the fact that both body and tarsi are equipped with hydrofuge hairs, keep it afloat. But if a drop of detergent is syringed onto the surface, the tension of the film is reduced and the pond skater will sink beneath the surface.

Most surface dwellers rely on the tension of the meniscus and upon hydrofuge hairs to support them. This is their habitat which they exploit to the full. *Gerris*, in common with other surface-dwelling invertebrates, possesses keen eyesight for locating any small insects falling on the surface. It swims quickly to the prey and seizes it with the short pair of front legs held above the water. The victim is then stabbed with the beak or rostrum, possessed by all the water bugs (Hemiptera), and the body juices are extracted. These methods are used by other members of the Hemiptera such as the water measurer (*Hydrometra stagnorum*) and the water cricket (*Velia currens*).

When trapped beneath the surface, a silvery covering can be seen which is a layer of air, enclosed by the hydrofuge hairs, surrounding the body.

The whirligig beetle, *Gyrinus natator*, is another surface dweller using the large fan of hairs on its third pair of legs (Fig. 11.1 (b)) to propel it in endless gyrations. This beetle has special modifications of the eyes which permit it to see both above and below the surface film. Approaching danger will cause it to dive swiftly, using its extra powers of propulsion to break

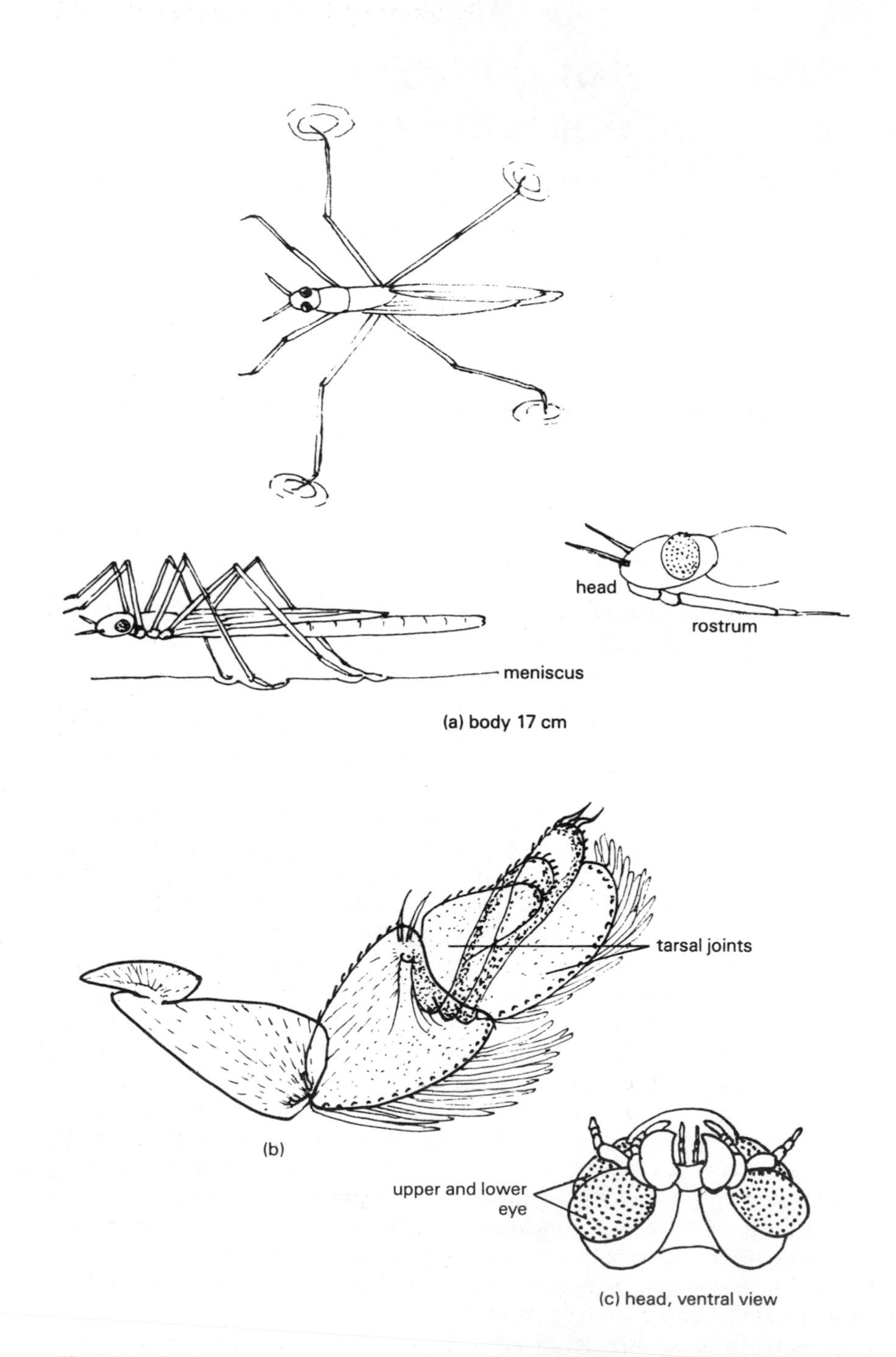

Fig. 11.1 Surface dwellers. (a) Pond skater, *Gerris* sp. The body is covered with unwettable (hydrofuge) hairs so that it is supported by the meniscus. Skating on the second and third pair of legs, the first pair are held in front of the head to catch prey falling on the surface. The rostrum is then extended to pierce and kill the prey. (b) Third leg of whirligig beetle, *Gyrinus natator.* The tarsal joints, opened out like a fan, provide an efficient swimming paddle. (c) The eye is double, one half capable of vision above the surface, the other beneath it

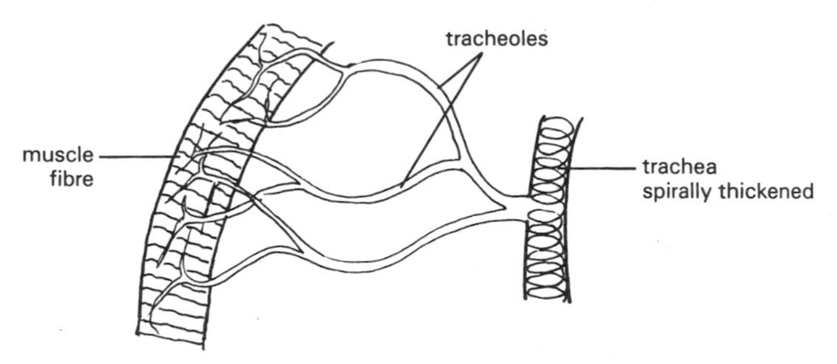

Fig. 11.2 Semi-diagrammatic representation of a portion of an insect's tracheal system

the meniscus. While on the surface, *Gyrinus* obtains its supply of air directly from the atmosphere. When it is forced to dive, air is carried underneath the wing cases and in a bubble attached to the end of the abdomen. This bubble is then used as a kind of lung, as in *Corixa* (p. 114).

Partially aquatic invertebrates

This group comprises animals which, although aquatic, must have access to atmospheric air for their supply of oxygen. In order to acquire this, various forces have to be overcome such as buoyancy and the ever present barrier between water and air, the surface film.

The majority of partially aquatic species are insects, some of which are aquatic for the whole of their life cycle. They employ various methods to obtain their supply of air.

Almost all insects breathe by means of a complicated tracheal system. This consists of a series of tubes which branch to supply every part of the body. Ultimately, the finest tubes come into intimate contact with the colourless blood where an exchange of oxygen and carbon dioxide takes place by diffusion (Fig. 11.2). The lumen of the tubes is kept open by spiral thickening of the walls with chitin. In many aquatic insects the tubes join up and open to the surface of the body through *spiracles*. The positioning of the spiracles varies in aquatic insects according to the method adopted for obtaining atmospheric air.

The larva and pupa of the mosquito, *Culex pipiens* (Fig. 11.3 (a, b)), are examples of partially aquatic forms in which the spiracles are in different positions. In the larva they open at the end of a special breathing siphon on the next to last segment of the abdomen. To obtain air the larva reaches the surface by a series of looping movements of the body. It then breaks the surface film by means of hydrofuge hairs on the last abdominal segment, and the siphon comes into contact with the atmosphere. Two large tracheal tubes traverse the siphon and continue along the entire length of the body, branching to supply each segment. Whilst at the surface, the larva feeds incessantly by means of the mouth brushes and the antennae which comb the water just beneath the surface for food particles, conveying these to the mouth. Breathing and feeding go on simultaneously, the two operations occurring at opposite ends of the body. To remain, often for long periods, in this almost vertical position, its body must be slightly denser than water, proof of this being that if disturbed from above, the larva sinks slowly to the bottom by gravity.

In the pupa of the mosquito (Fig. 11.3 (b)), the spiracles open at the end of a pair of breathing trumpets on the head. The reversal of the respiratory arrangements in the pupa is due to the fact that the imago, which is winged and aerial, emerges from a split in the thorax. This makes it more convenient for the spiracles to be in the head region.

The respiratory habits of both the larva and pupa of the mosquito afford an easy method of control by the aerial spraying of bodies of water with oil. The oily film prevents respiration

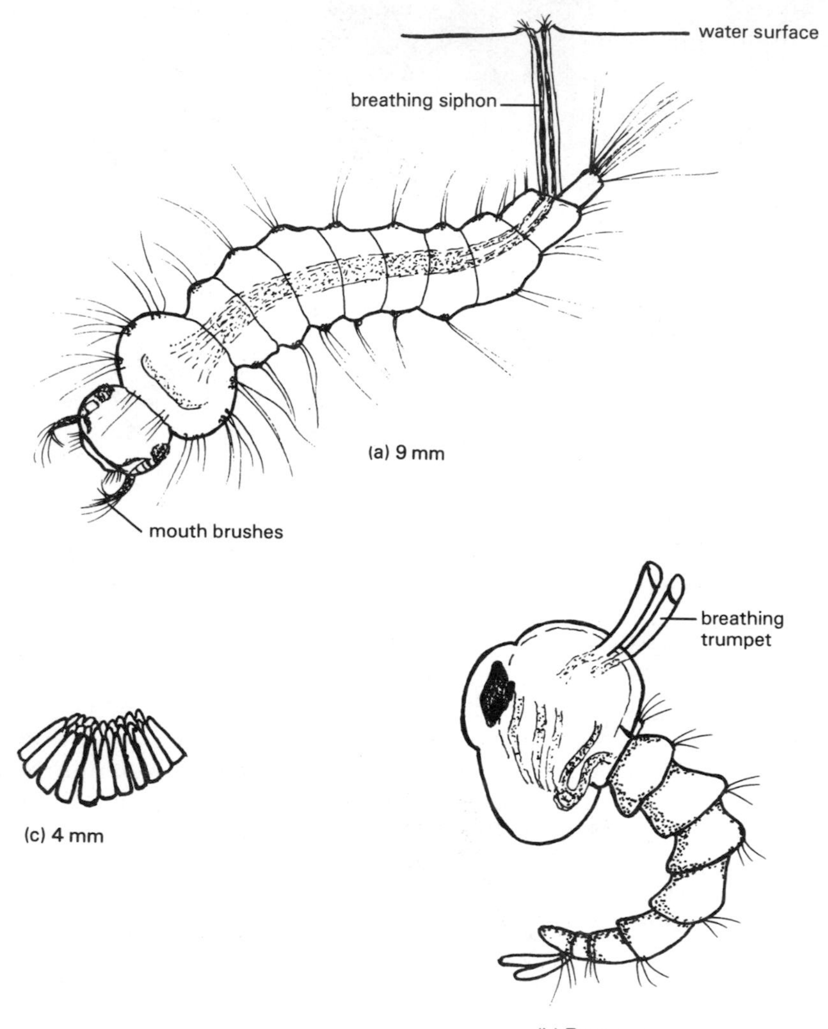

Fig. 11.3 Mosquito, *Culex pipiens.* (a) Larva in breathing position at the surface. (b) Pupa. (c) Egg raft floating at the surface, consists of 250 or more cigar-shaped eggs glued together. The larva escapes through the lower end of the egg into the water

taking place. This method is used to eliminate the malarial mosquito, *Anopheles*, whose larvae and pupae respire in the same manner as *Culex*.

Another two-winged fly, the soldier fly, *Stratiomys* sp. (Fig. 11.4), breathes as a larva in a similar way to the mosquito but instead of being borne on a siphon, the spiracles open at the end of the abdomen and are surrounded by an intricate coronet of filaments. These filaments fan out when the larva reaches the surface tail first. They support the animal which is hanging head down. Its body can extend to 3 cm or more and while hanging from the surface, it feeds like *Culex* on minute organisms. *Stratiomys* pupates within the larval skin. At this time the larva often leaves the water to bury itself in mud or amongst floating vegetation.

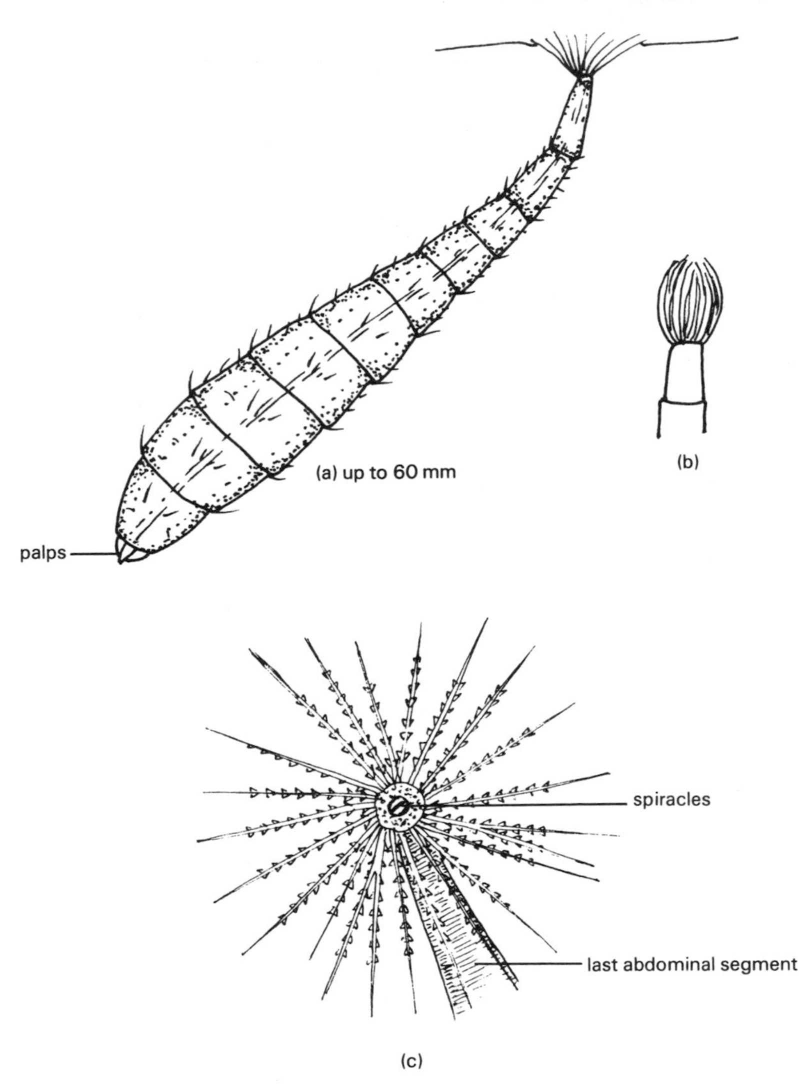

Fig. 11.4 (a) Larva of the soldier fly, *Stratiomys* sp., breathing at the surface. (b) Tail filaments drawn together to enclose a bubble of air when beneath the surface. (c) Coronet of tail filaments surrounding central spiracles

A third species of two-winged fly, *Eristalis* sp., the rat-tailed maggot, also passes its egg, larval and pupal stages in fresh water. A description of this extraordinary larva (Fig. 11.5) is necessary in order to show the extreme modification it has undergone. It is admirably adapted to life at the bottom while respiring atmospheric air. The larva lives in the muddy sediments of shallow pools and solves the problems of reaching the surface to obtain air in the most ingenious fashion. The body lies prone in the mud but the last segment is formed into a siphon inside which lies another tube and inside that yet another. If the water becomes deeper the tubes are extended like the sections of a telescope and at their fullest can reach several centimetres (but see Réaumur's observation on p. 111) to bring the circlet of eight

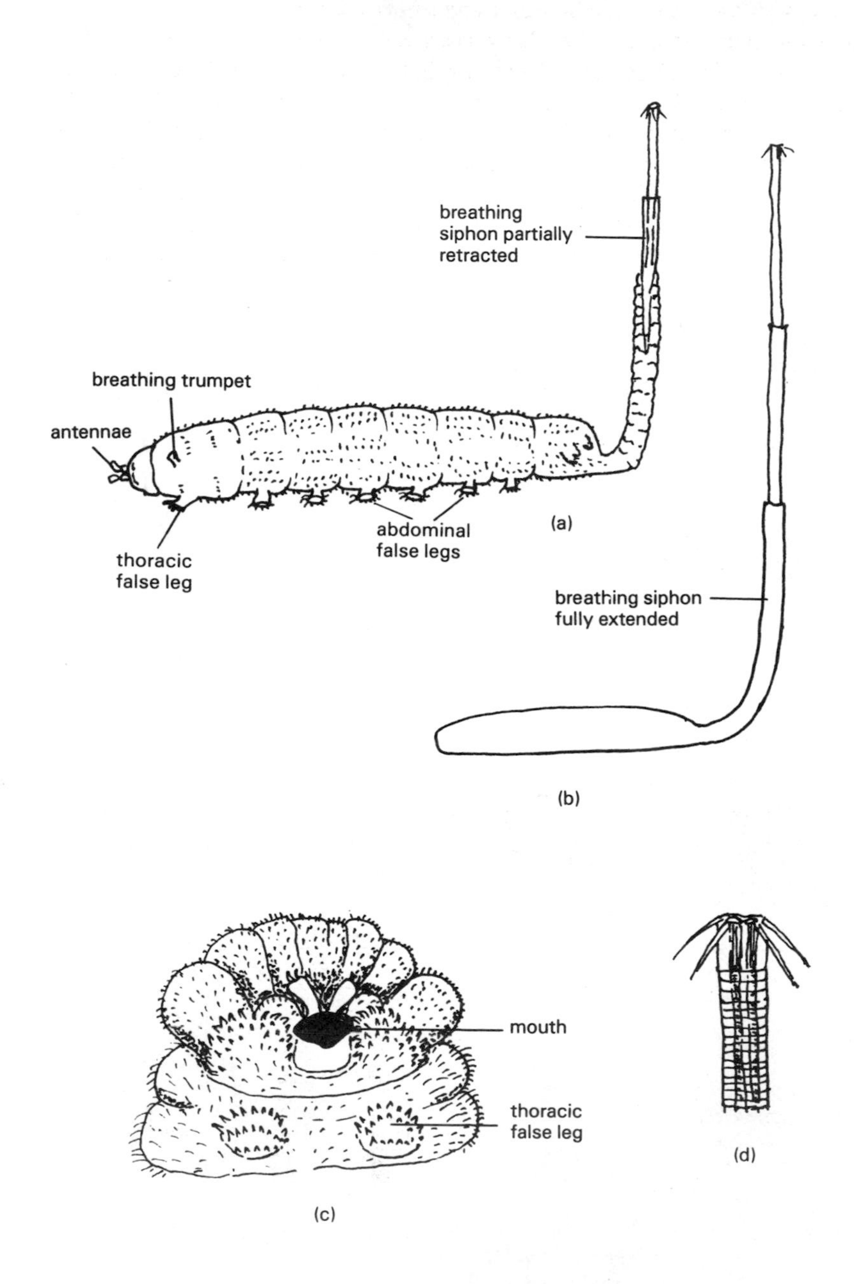

Fig. 11.5 Larva of rat-tailed maggot, *Eristalis* sp. Body 15 mm. (a) Lying in mud in shallow water. (b) Breathing siphon extended to reach surface in deeper water. (c) Ventral view of head, showing mouth surrounded by a number of mobile lobes equipped with small hooks and hairs. In feeding, the lobes push food particles towards the mouth. (d) Eight stout hairs at the tip of the breathing siphon fold together to enclose a bubble of air when the siphon is retracted beneath the surface

feathery hairs to the surface. The spiracles lie in the centre of the circlet and when the siphon is withdrawn, the hairs are drawn together enclosing a bubble of air.

A lengthy and fascinating account of this larva was given by Miall [11.1] who records a description made by the French scientist, Réaumur, in the early eighteenth century. One of Réaumur's penetrating observations is worth recording. Quoting from Miall: 'To ascertain whether they (the larvae) could lengthen them (the siphons) yet more, Réaumur added water to the depth of half an inch, when the tails were lengthened to the same amount. Again and again water was added, and the tails became lengthened in proportion. When however, the depth grew to five and a half or six inches, the larvae could no longer reach the surface from the bottom'. Another early worker, Wilkinson, of whom we have no record, says

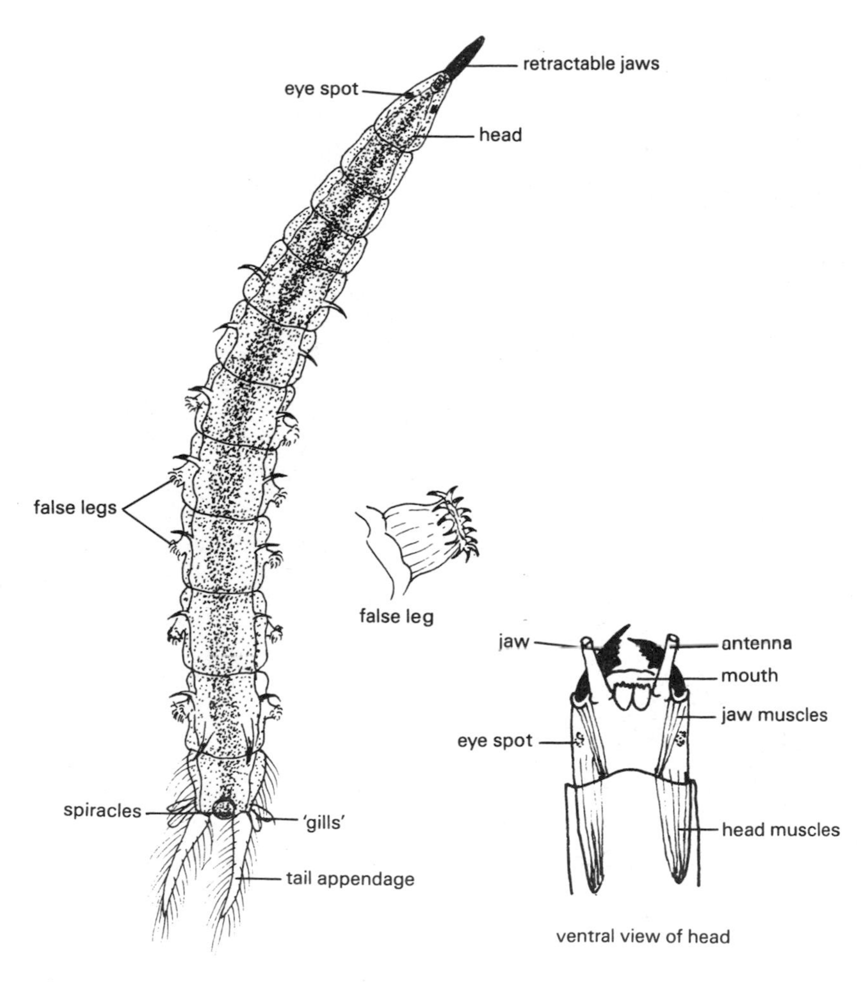

Fig. 11.6 Cranefly larva, *Dicranota* sp. found in muddy pools and streams. It moves about actively in the mud by means of its false legs, provided with a circlet of hooks, in search of its prey, *Tubifex* (Fig. 11.17). The head is equipped with powerful muscles to retract the jaws, and the head itself, within the body

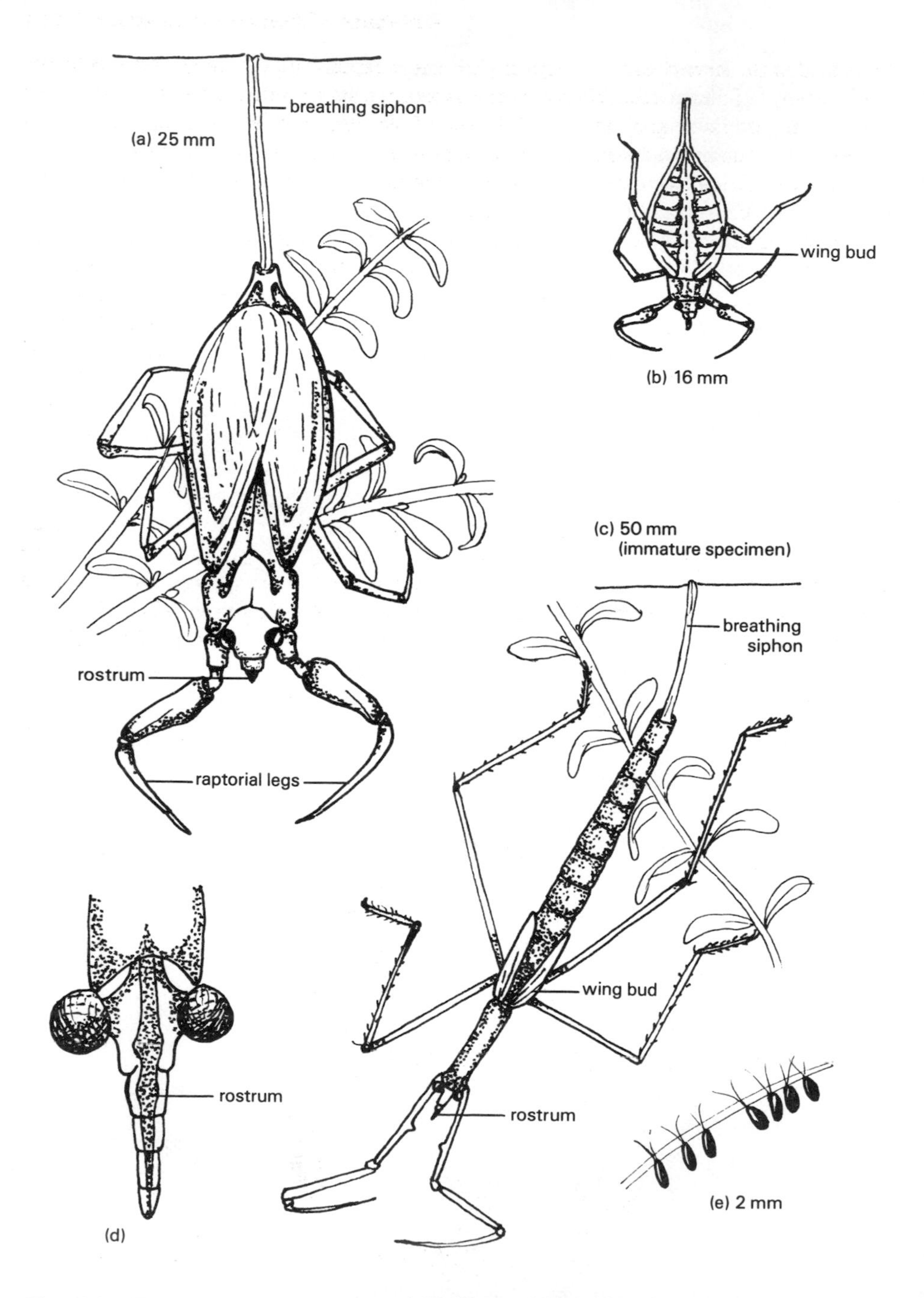

Fig. 11.7 Two freshwater bugs (Hemiptera). (a) Water scorpion, *Nepa cinerea.* Note the long, double breathing siphon. The hard, black wingcases cover purplish, membranous wings. (b) Nymph of *N. cinerea*, showing the developing siphon and wings. (c) Water stick insect, *Ranatra linearis*, immature specimen with developing wing buds. Note the rounded body and very long legs which, when aligned with the body, make it difficult to see. (d) Head of *R. linearis.* Both *R. linearis* and *N. cinerea* possess a rostrum which pierces the prey and sucks its juices. (e) Eggs of *R. linearis* have two appendages. The eggs are inserted individually into incisions made in submerged plant material

'The food of the larva consists of organic particles, scraped from submerged objects by the hooklets around the mouth. When the larva is feeding, its movements resemble those of a pig, working over a heap of refuse with its snout'. We owe much to these early naturalists whose detailed observations add a significant dimension to our modern knowledge.

The larva of *Eristalis* pupates within the larval skin and the puparium then floats passively to the surface where breathing trumpets function to take in air.

Mention must be made of the larva of a species of cranefly, *Dicranota* (Fig. 11.6), which lives in muddy pools and slow-flowing streams in which dwell its prey, *Tubifex* worms (Fig. 11.17(b)) and p. 125). *Dicranota* travels through the mud by means of false legs equipped with a circlet of hooks, pursuing *Tubifex* into their burrows. It feeds on the worms using its strong jaws which can be retracted into the head which, in turn, can be completely retracted into the thorax. The tail appendages are thin-walled and permeable to gases. Two pairs of 'gills' are also present. Thus *Dicranota* can obtain its supply of oxygen when under water by means of these 'gills' but can also breathe air at the surface by protruding the terminal spiracles above the water.

The nymphs and adults of the water scorpion, *Nepa cinerea*, also possess breathing siphons (Fig. 11.7 (a,b)) but unlike *Eristalis* cannot extend them. Being a poor swimmer, *Nepa* must crawl up weeds to the surface and turn head downwards. Hydrofuge hairs at the tip of the siphon break the surface film and air can be drawn into the spiracles without water entering the tracheal system. A near relative is the water stick insect, *Ranatra linearis*, (Fig. 11.7(c,d)), which breathes in the same way. Both *Nepa* and *Ranatra* lay their eggs individually in incisions made in submerged plant stems and leaves (Fig. 11.7 (e)). Each egg bears appendages which are probably respiratory in function.

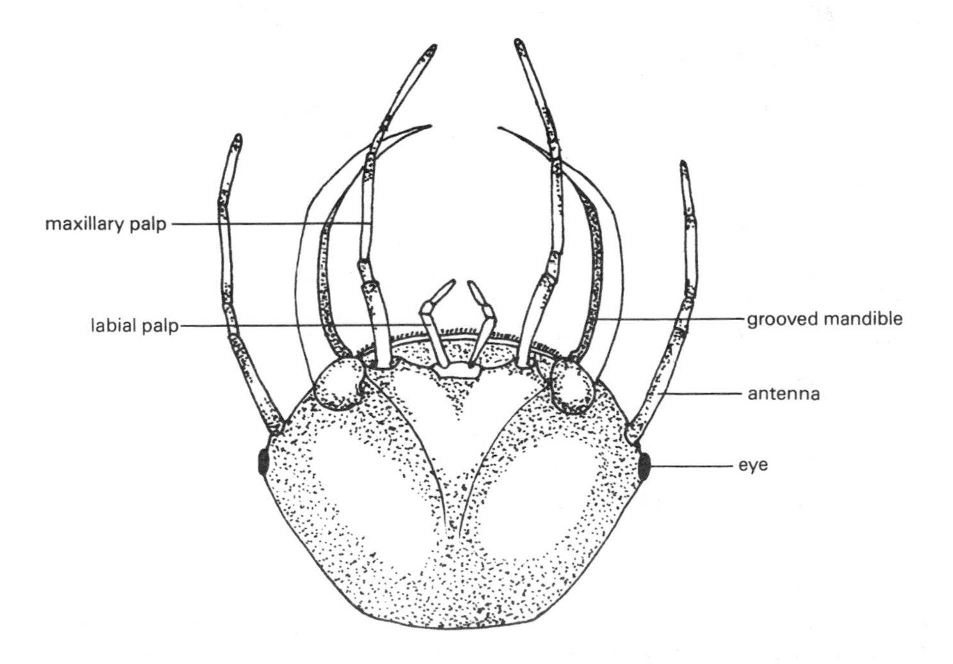

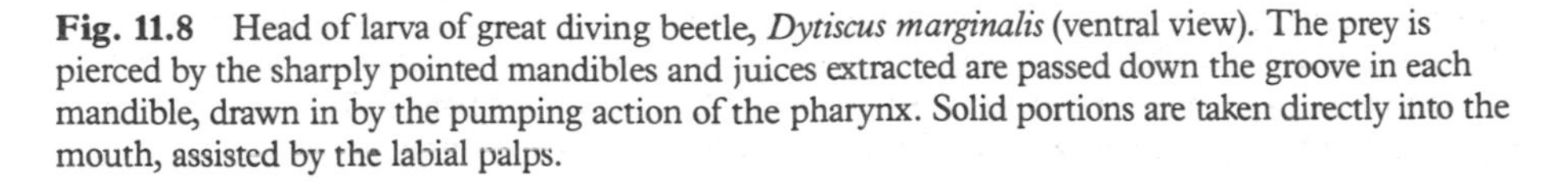

Fig. 11.8 Head of larva of great diving beetle, *Dytiscus marginalis* (ventral view). The prey is pierced by the sharply pointed mandibles and juices extracted are passed down the groove in each mandible, drawn in by the pumping action of the pharynx. Solid portions are taken directly into the mouth, assisted by the labial palps.

A similar method of respiration is employed by the larva of the great diving beetle, *Dytiscus marginalis* (Fig. 6.4 (a)), which rises tail first to the surface by looping the body, assisted by 'walking' movements of the fringed legs. The larva is a voracious carnivore and, after catching its prey, hangs from the surface taking in air while at the same time sucking the juices of the prey by means of its hollow, hooked mandibles (Fig. 11.8). The adult beetle (Fig. 6.4 (b)), also carnivorous, takes in air in the same manner as its larva, air passing beneath the elytra to paired spiracles on each abdominal segment.

The water boatman, *Notonecta glauca*, has other problems for, when carrying air, its body is lighter than the surrounding medium. To remain beneath the surface it must either cling to submerged weeds or keep swimming head downwards with its powerful third pair of legs. To reach the surface it releases its hold and floats up. Again, like many partially aquatic animals, it uses a brush of hydrofuge hairs surrounding the tip of the abdomen to break the surface film. Air is drawn in and held as a bubble beneath the abdomen which is clothed in long hairs (Fig. 11.9). Once this has taken place its body becomes more buoyant so that much energy is expended in keeping it beneath the surface, where it seeks its food consisting of other small insects, worms, and even small tadpoles.

A close relative, *Corixa*, sometimes called the lesser water boatman, feeds on detritus and small organic particles in sediment at the bottom. Every now and then it must rise to the surface to replenish its air supply carried as a bubble at the tip of the abdomen and it must maintain constant swimming movements to overcome buoyancy. The bubble acts as a kind of 'lung' inasmuch as oxygen will enter the bubble by diffusion when the concentration of oxygen within it falls below that of the surrounding water. This physiological process permits the insect to remain beneath the surface and within its feeding region for long periods. Doubtless the ventral air bubble carried by *Notonecta*, and that of other comparable species, functions in much the same manner.

All partially aquatic beetles must adopt some method of obtaining air from the atmosphere but the great silver water beetle, *Hydrophilus piceus* (Plate 11.1) manages this in an extraordinary way. Being a poor swimmer, it rises to the surface by 'walking' movements of its fringed legs. Once there, it assumes a nearly horizontal position, bringing the junction between the thorax and the head on one side of the body to the surface, and coiling the antenna round the side of the head (Fig. 11.10 (a)). Hydrofuge hairs surrounding the last

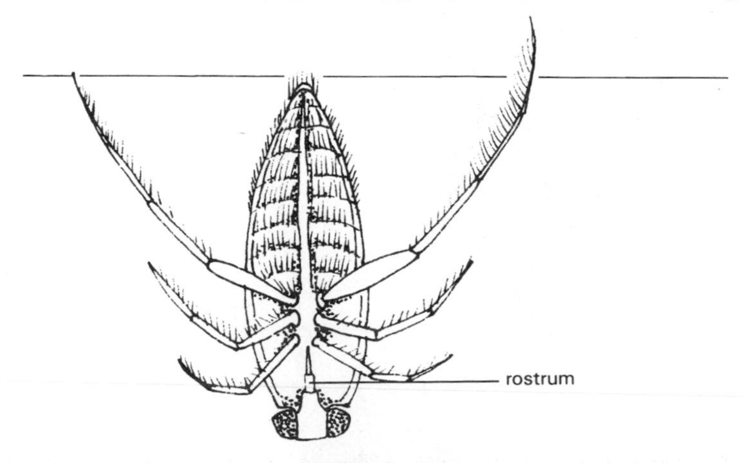

Fig. 11.9 Water boatman, *Notonecta glauca* (ventral view), taking in air at the water surface. Hydrofuge hairs at the tip of the abdomen break the surface and air drawn in is stored beneath the thick covering of hairs beneath the abdomen. Prey is killed by poison injected by means of the rostrum, as with other hemipterans. Only liquid juices are extracted

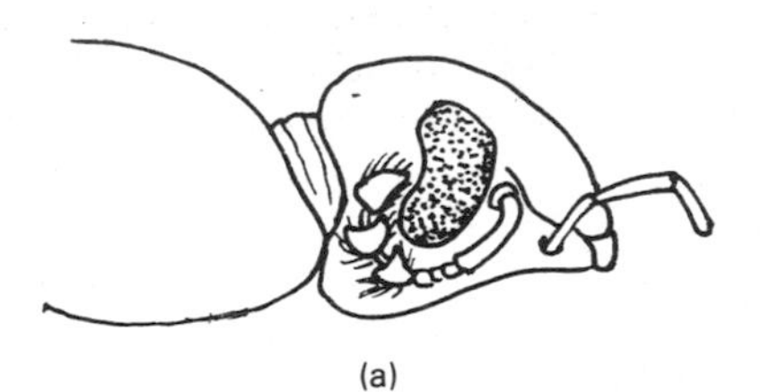

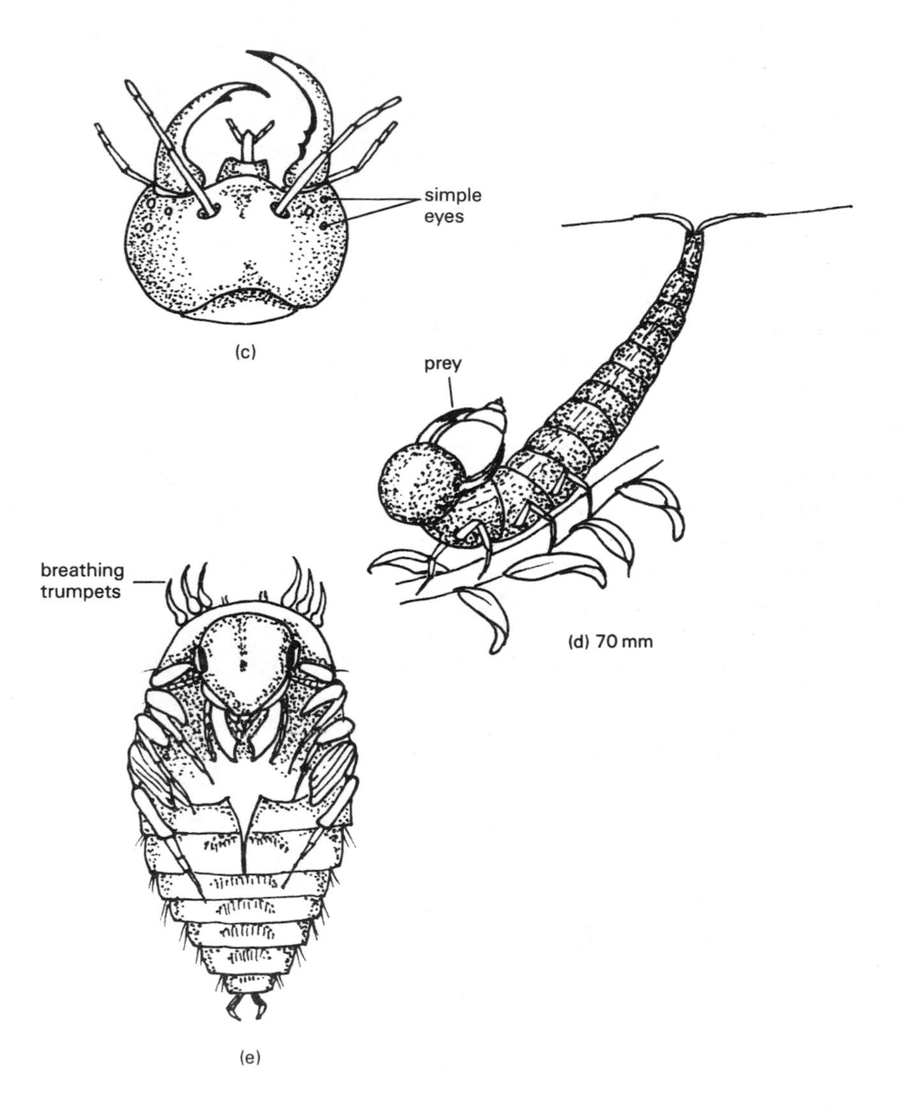

Fig. 11.10 Great silver water beetle, *Hydrophilus piceus*. (a) Head of beetle showing antenna coiled round the eye when the insect is taking in air at the surface. (b) Shell of a mollusc eaten by larva of *H. piceus*. (c) Head of larva. The right mandible is larger than the left. (d) Fully grown larva with head thrown back and clasping prey, feeding and at the same time respiring at the surface. (e) Pupa showing breathing trumpets

Plate 11.1 Male great silver water beetle, *Hydrophilus piceus*. 36 mm long. Note the expansions on the forelegs which are absent in the female. *H. piceus* is the second largest British beetle

antennal segments break the surface film and air is drawn in and expelled by movement of the abdomen. Fresh air fills the cavities beneath the elytra, which are in direct contact with the spiracles, and also covers the region beneath the abdomen which is thickly clothed with hairs. Equipped with a fresh air supply, the beetle can remain beneath the surface for many minutes.

H. piceus [11.3] is now very local in distribution, being confined mostly to the Somerset Levels [11.4]. Mating takes place in May or early June and soon afterwards the female builds a large egg cocoon (Plate 11.2) from silk spun from spinnerets at the end of the abdomen. Eggs are laid simultaneously with the construction of the cocoon, usually about 60 in number, and after eleven days they hatch into larvae which feed at first on rotifers, diatoms, and other small organisms on pondweed leaves. Soon they assume a totally carnivorous diet consisting mostly of small planorbids and other molluscs (Fig. 11.10 (b)). Their strong mandibles (Fig. 11.10 (c)) are admirably constructed for holding the prey and crushing snail shells in order to extract the soft parts. The typical feeding position adopted by the larvae is shown in Fig. 11.10 (d). The mandibles, like those of the larvae of *Dytiscus*, are hollow and, after piercing the prey, digestive juices flow down them. Extra-digestion of the prey then takes place, the liquid food being taken into the mouth. After four ecdyses the larva pupates, building a chamber in soft peat or mud. Like the pupa of *Eristalis*, it is equipped with a pair of breathing trumpets which supply it with sufficient air for its now much reduced needs (Fig. 11.10 (e)).

The four stages of the life history of *H. piceus* illustrate the different modifications of structure to suit the requirements of each stage. The female beetle will only construct an egg cocoon, made of floating fragments of weed, in a ditch in which there is an abundance of molluscan food for the carnivorous larvae. The soft peat of the Somerset Levels offers ideal conditions for pupation and its well-weeded habitat in late summer provides an abundance of food for the vegetarian adult beetles. There is fairly strong evidence that the adult beetles undertake quite long aerial migrations at night and it is possible that the Sedgmoor population is replenished with immigrants from across the English Channel.

Plate 11.2 Silken egg cocoon of *H. piceus*. 20 mm long. Note the fragments of weed attached to the cocoon to give it protection and buoyancy. The mast, constructed of a different silk, is thought to afford a passage for air to the eggs inside

Although several species of spider live near, or even on, fresh water, *Argyroneta aquatica* (Plate 11.3) is the only one which lives permanently below the surface. This spider is possibly the best example of a partially aquatic invertebrate which can exist for long periods without having to make a journey to the surface for replenishment of oxygen. The whole body of *Argyroneta* is covered with long hairs in which a single bubble of air is trapped. Both male and female spiders construct a silken bell amongst weed and, making many journeys to the

Plate 11.3 Water spider, *Argyroneta aquatica*. Note the silvery air bubble surrounding the body (Photo: Andrew W. Cooper)

surface, descend after each visit with an air bubble which is rubbed off with the hind legs, under the bell. Gradually the bell is filled with air in this way and the spider has a plentiful supply. Several bells may be constructed and one will be used by the female in which to lay her eggs. The young spiders, when they hatch, do not make their own shelters at first but use small empty snail shells which they fill with air.

Some freshwater molluscs (pulmonates) take in air through a respiratory aperture and must return to the surface frequently to replenish supplies. The air circulates in the mantle cavity where gas exchange takes place through the highly vascular walls of the mantle. In others, dissolved oxygen in water circulating in the mantle cavity is used for respiratory purposes, enabling them to be independent of depth.

The wandering pond snail, *Limnaea pereger* (Fig. 9.3 (d)), exists in two forms, using both methods of respiration. Populations of each kind are often to be found in the same area, one in deeper water, using an aqueous lung, and the other with an aerial lung in shallower conditions. This species may well be an example of an evolutionary trend from water to land.

The great pond snail, *Limnaea stagnalis*, breathes air and spends much of its time at or near the surface, using the surface film for gliding along on its expanded foot (Fig. 11.11).

Totally aquatic invertebrates and fish

To extract dissolved oxygen from the water, all totally aquatic animals must either possess gills of one sort or another or else have a body covering through which gaseous exchange is possible. The latter means that the animal usually has a large surface area: volume ratio so that an exchange of gases can be maintained without resort to respiratory organs.

This method of gaseous exchange usually involves a thin and highly permeable skin. Flatworms (Figs. 9.2 and 10.4, p. 75), often found in large numbers gliding over submerged vegetation or under stones, breathe in this way. They are carnivorous, feeding on many kinds of small insects or crustaceans, alive or dead. This they do by protruding a pharynx from the ventral surface of the body. If the prey is small it is taken whole but if it is larger it is wrapped in slime and pieces are sucked off. Their essential respiratory and nutritional needs are met in situations varying from the surface of underwater leaves to stony or muddy substrates, so it is not surprising that flatworms are ubiquitous.

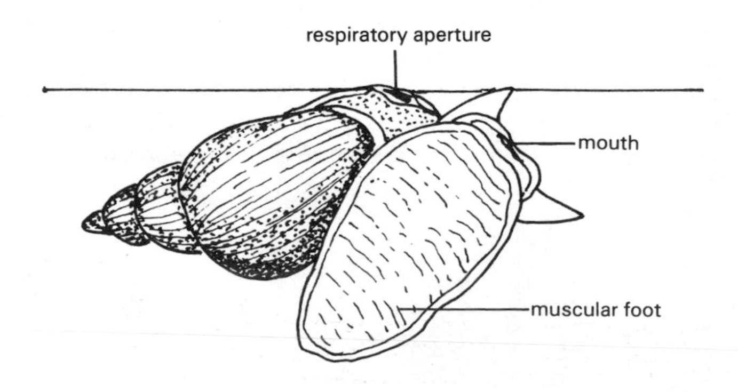

Fig. 11.11 Great pond snail, *Limnaea stagnalis*, taking in air at the surface. The air fills the shell cavity, lined with the mantle which has a rich blood supply. The mantle acts as a 'lung'. Soft parts of the body, in contact with the water, are also probably able to effect gaseous exchanges.

Some of the larger freshwater invertebrates, such as leeches and other annelids which have a low metabolic rate, are also able to maintain an exchange of gases through the body wall without special respiratory structures. Smaller crustaceans, too, such as water fleas and ostracods, fall into this category for, although they have a hard exoskeleton, their limbs are modified in various ways to create a current of water which flows over the soft parts of the body. This also acts as a feeding current by wafting particles of food towards the mouth (Fig. 3.3).

Many Diptera have totally aquatic larvae and like the animals mentioned above, have no special respiratory organs. One example is the predatory larva of the phantom midge, *Chaoborus* sp. (Fig. 11.12), which besides being of interest because of its strange habits and structure plays an important part in the food relationships of still-water communities. It may be present in considerable numbers, especially in pools overhung by trees, but because of its transparency, is extremely difficult to see. The larva can remain for a long time lying horizontally beneath the surface on the look-out for its prey.

Food consists of zooplankton such as crustaceans, other fly larvae, and rotifers, all of which are grasped by the antennae, modified as prehensile organs. A tail fin, composed of a comb of bristles, acts as a rudder and assists in propelling the larva through the water as it progresses with jerky side-to-side movements of the body.

In *Chaoborus* the chitinous body covering is so thin that direct gaseous exchange can take place. Air also fills the bean-shaped sacs at either end of the body. These are hydrostatic organs which adjust the buoyancy of the larva to the density and pressure of the water by expansion or contraction, thus permitting it to rise or fall. In this way different regions can be explored for food as it moves between the bottom sediments and the epilimnion above [11.2].

The adults of *Chaoborus* lay eggs in summer and the first instar larvae dwell amongst the plankton on which they feed. The second and third instars are quickly passed and spend their time partly in the plankton and partly amongst the sediments. The fourth and final instars emerge in the spring and migrate to feed on the abundant zooplankton. They pupate near the surface and the imagos emerge in July. The life history therefore takes about twelve months to complete during which time the food preferences at different life stages are catered for by the ability to migrate.

Various kinds of gill are possessed by animals dwelling totally below the surface of fresh water. Like the many partially aquatic animals, their mode of respiration is closely linked with obtaining food, often involving the creation of a current of water passing over the gills which at the same wafts in particles of food.

Basically a gill is an outgrowth from the body through the surface of which diffusion of oxygen and carbon dioxide can take place, similar to the way in which these gases diffuse through the moist lining of the lungs in a mammal to reach the bloodstream.

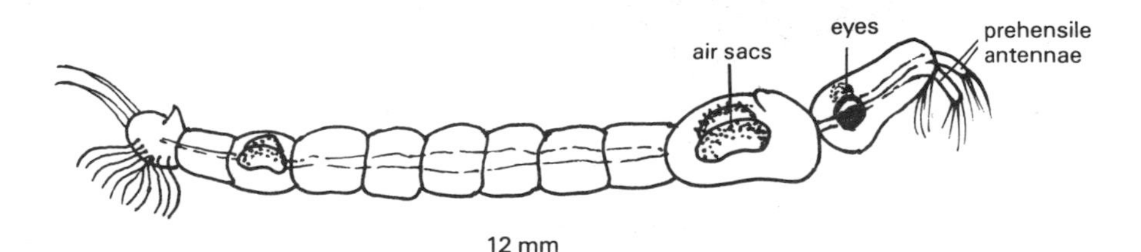

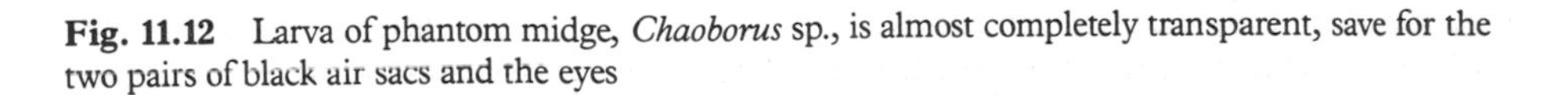

Fig. 11.12 Larva of phantom midge, *Chaoborus* sp., is almost completely transparent, save for the two pairs of black air sacs and the eyes

The nymph of the mayfly, *Chloeon dipterum*, typical of still water, is to be found at most times of the year (Fig. 11.13). The abdomen bears pairs of plate-like gills and each is double, except the last. The first six pairs are kept in constant vibration, setting up a current of water which passes backwards until it meets the seventh pair of gills. These are kept stationary and deflect the current to each side so that a fresh supply of water is brought into use. If the water is highly oxygenated, movement of the gills is slow. But if the nymphs are transferred to water which has been boiled and then cooled to the original temperature, the gills immediately start to vibrate much more rapidly in an endeavour to waft oxygen towards them.

Nymphs of damselflies (Fig. 11.14 (a)) have three flattened laminate gills on the tail which operate in much the same way as those of *Chloeon* to create a current flowing over them. These lamellae contain a number of tracheae and are usually considered to function as gills. However, if they are accidentally lost, the nymph is apparently able to breathe perfectly well through the body wall. Damselflies are usually found amongst pondweeds where the oxygen concentration of the water is high due to photosynthesis. The plate-like gills undoubtedly help in locomotion when these active nymphs move about in search of prey with backwards and forwards movements of the abdomen.

Dragonflies all have aquatic nymphs but in contrast to damselflies, the sluggish nymphs of the large hawker dragonflies (Fig. 11.14 (d)) have internal gills lining the rectum. Water is pumped in and out of the anus by alternate contraction and expansion of the abdomen. This creates a constant current through the gill chamber. It is also a means of rapid movement, for as water is expelled from the anus, the nymph is jet-propelled forwards. Normally living in the bottom sediments of still and slow-flowing water, they are dull shades of brown, resembling their background. The nymphs can remain motionless for long periods, relying on stealth and good eyesight to stalk prey. This they seize by shooting out the mask which, when not in use, lies folded beneath the head (Fig. 11.14 (c)). At the end are strong, hooked jaws which grasp the prey. The whole mask is then pulled back to bring the food to the mouth. Damselfly nymphs also possess these mouth parts which are more slender but are used in the same way.

Dragonfly nymphs may remain up to two years in the water before they are ready to emerge as winged adults. Emergence from the nymphal skin must take place above the water, so the mature nymph crawls up a plant stem. After emergence, the adult remains for an hour or so clinging to the skin while its wings stretch and dry. Soon the adults are ready to mate and the female then lays eggs on a water-plant stem, usually just beneath the surface, and the life cycle

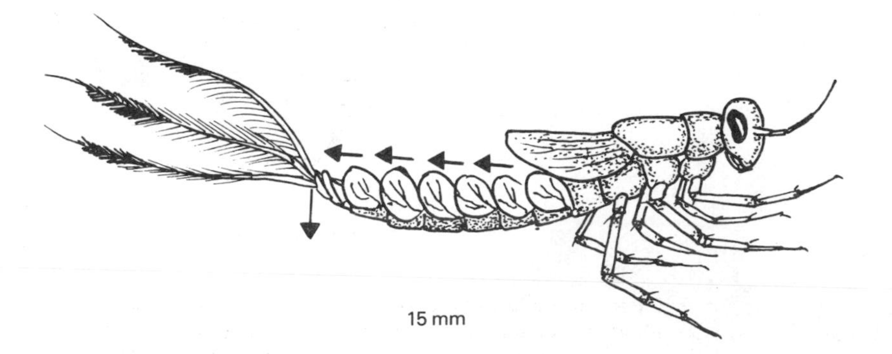

Fig. 11.13 Nymph of a mayfly, *Chloeon dipterum* (Ephemeroptera), typical of still water. A current of water created by the constant movement of six pairs of abdominal gills is deflected by the seventh pair so that it will not be immediately used again

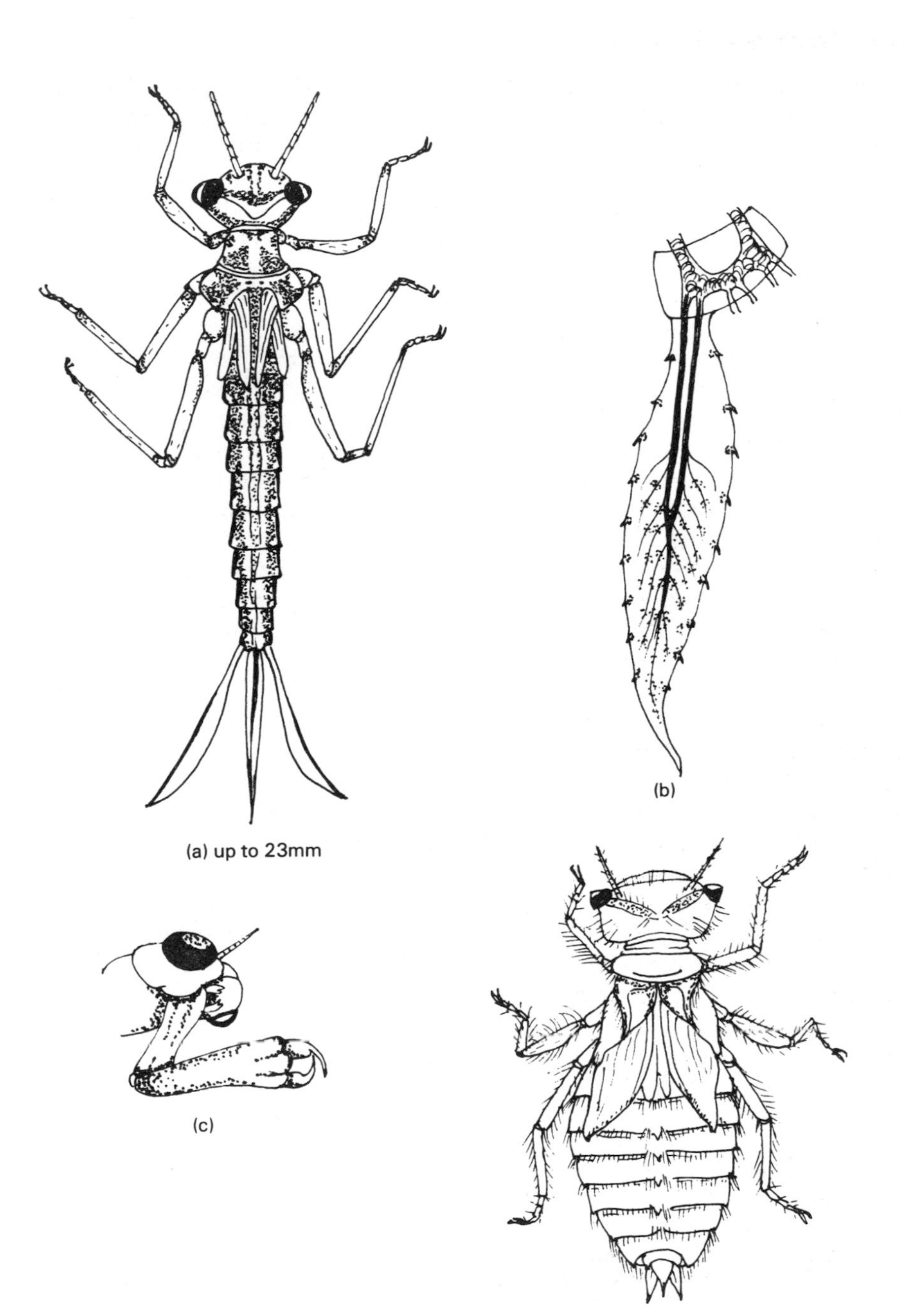

Fig. 11.14 Aquatic nymphs of dragonflies (Odonata). (a) *Ischnura* sp. Nymphs of all species of damselflies bear three leaf-like gills at the extremity of the abdomen. Their long legs assist in climbing amongst weeds where their green/brown colouring makes them inconspicuous. (b) Laminate gill of *Ischnura*, to show arrangement of the tracheal branches, a diagnostic character in distinguishing species. (c) The hinged mask in aeshnid nymphs can be extended some distance in front of the head. Both aeshnid and libellulid nymphs (d) possess rectal gills and the shape of the mask, possessed by all dragonfly nymphs, is also a diagnostic characteristic

is ready to begin once more. Like the mayflies, stoneflies and caddisflies, dragonflies are dependent on fresh water for all stages in their life history and, even as adults, their hunting grounds are usually close to or over the water. The hawker dragonflies in particular are strong fliers, catching their prey on the wing. They can migrate for long distances, thus ensuring dispersal of the species for colonization of fresh habitats.

Caddisflies (Order Trichoptera) pass through four life stages: the eggs, laid in water, hatch into aquatic larvae which pupate in water and emerge as winged adults.

The majority of trichopterans breathe by means of gills of one kind or another and the mandibles are modified according to whether they are carnivorous, herbivorous or mixed feeders. Many larvae build elaborate cases inside which they live (Fig. 11.15). Constant undulations of the body create a current of water which passes through the case and over the gills. Larvae living in well-weeded ponds and ditches use fragments of weed for case building, others use small bits of stick, sand, stones, and even the dead shells of molluscs. In some species the material of which the case is built and its shape can be diagnostic, but examination of the larva itself is necessary for accurate identification. Those species typical of streams and rivers are the caseless caddises. This can be a misnomer, for although some, like *Rhyacophila*, build no case at all, others construct elaborate cases during the latter part of their larval lives and many build simple cases of some sort. They must all attach themselves in some way to the substratum in order to avoid being washed away. *Hydropsyche* (Fig. 10.3) builds a rough shelter of stones, sometimes incorporating quite large pebbles, attached by strands of silk to a larger stone. A silk net is woven across the upstream opening to the shelter and long strands of silk mesh extend beyond this. Detritus and particles of food are caught in the mesh. Indeed, the different methods used by net-building species for trapping food makes an interesting study.

In contrast to the pulmonate molluscs (p. 118) there are some freshwater species which close the opening to the shell with a horny plate called the operculum, attached to the foot. The operculum fits the opening to the shell and when the body of the animal is withdrawn, closes it completely. These snails are called **operculates** and are found in reasonably well-aerated water, breathing dissolved oxygen by means of special gills. In the common valve snail, *Valvata piscinalis* (Fig. 11.16 (a)), the gill is particularly evident.

A more advanced type of gill is found in the freshwater bivalve molluscs. That of the swan mussel, *Anodonta cygnea*, is clothed with cilia which, by their constant beating action, pass

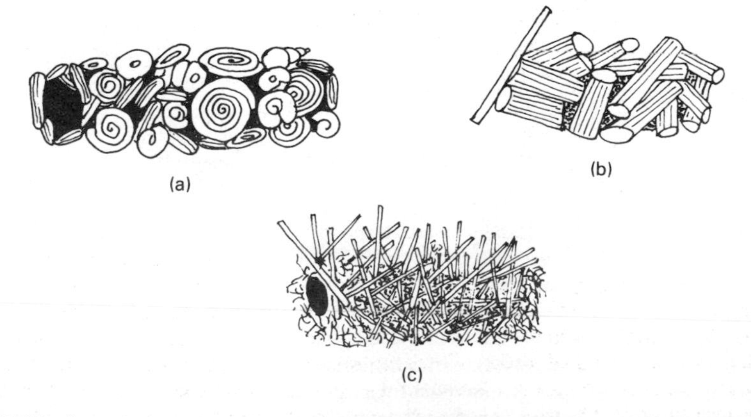

Fig. 11.15 Larval cases of some caddises. Materials readily available are used by larvae living in well-weeded waters. (a) Case composed of small mollusc shells and seeds of grasses. (b) Short lengths of reed have been used here. (c) Small fragments of sticks and roots have been bound together

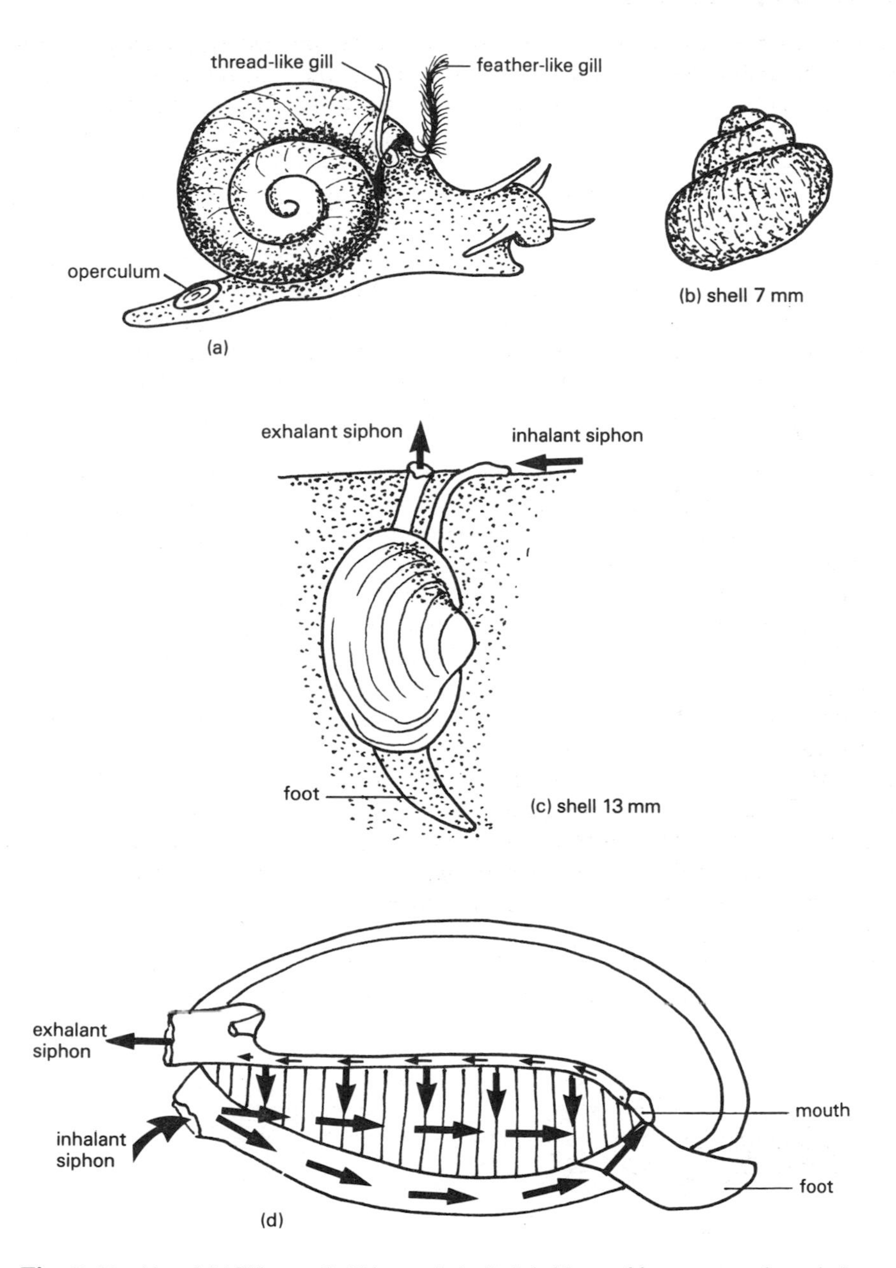

Fig. 11.16 (a) and (b) Valve snail, *Valvata piscinalis*, inhabits muddy streams and ponds, its feathery gill waving about. When withdrawn into its shell, as with other operculate gastropods, the entrance to the shell is closed by the operculum. (c) Orb cockle, *Sphaerium* sp. lying partially submerged in mud. The inhalant siphon sweeps the surface of the mud, taking in edible particles with the water. After circulating in the gill chamber, the water passes out through the exhalant siphon. (d) Schematic longitudinal section of the swan mussel, *Anodonta cygnea*, with the right lobe of the mantle removed to show the outer lamella of the right gill. Water taken in at the inhalant siphon passes over the gills. The vertical arrows represent ciliary currents passing down the gill, those at the bottom of the gill indicate the direction of the main food current running towards the mouth, while those at the top of the gill indicate the direction of the current passing to the exhalant siphon

a string of mucus, containing food particles, up the gill filaments and to the mouth. Here the gill performs both a feeding and respiratory role (Fig. 11.16 (d)) for water is drawn in through the inhalant siphon and passes out through the exhalant siphon, having circulated around the gill cavity.

A similar system is employed by the orb cockle shells, *Sphaerium* spp. (Fig. 11.16 (c)), which are commonly found in the bottom substrates of ponds and slow-flowing water. They move over the surface of mud by means of an extensible worm-like foot but can live partially submerged so long as the inhalant and exhalant siphons can reach the surface. Vacuum-cleaner-like movements draw in through the inhalant siphon a stream of water containing particles of mud and food. Since calcium is necessary for shell-building, molluscs thrive best in alkaline water with a fairly high calcium content, as we saw earlier (p. 10). Many aquatic plants also thrive in calcium-rich water and provide food for pulmonate molluscs. The presence of lime in the water is especially important because it causes precipitation of clay particles which would otherwise be free to clog the delicate respiratory mechanism of these molluscs.

Freshwater fishes possess highly developed gills, five pairs in all, protected by a horny operculum (Fig. 10.10 (c)). The gills are well supplied with blood containing haemoglobin. Water drawn in through the mouth is forced as a current over the gills when the mouth is closed, gaseous exchange taking place at it flows over the gill filaments. Some fish, such as the carp, actively gulp air at the surface and this is swallowed. In these fish the intestinal walls are richly vascular so that oxygen from the swallowed air is absorbed.

Inhabitants of silt and mud

The zones of water described so far have been those of the surface and open water. But at the bottom, among sediment, the oxygen content of the surrounding water can be greatly reduced by the accumulation of decaying vegetation and animal matter. Even under such conditions of anoxia there are some animals which are capable of exploiting this habitat.

The larvae of chironomid midges, of which there are many species, play an important part in the mud/silt habitat. Species with colourless or green larvae are usually found among floating vegetation or algae growing on stones in running water where dissolved oxygen levels are relatively high. But there are others which live in mud tubes made of silt. A scoopful of mud from the bottom of a stagnant pool will almost certainly contain a number of these larvae, easily recognized by their red colour. This is due to the presence in the blood of haemoglobin which is capable of combining with oxygen to form oxyhaemoglobin. In this way the oxygen-carrying powers of the blood are enormously increased. Nevertheless, it has been calculated that the volume of haemoglobin contained in a body the size of a chironomid larva could not store more oxygen than would last approximately twelve minutes without replenishment.

The adult chironomid midges lay their eggs in jelly masses on the surface of the water and the semi-planktonic larvae, which hatch from the eggs, drift over soft mud. Here they construct U-shaped burrows opening at each end at the mud surface (Fig. 11.17 (a)). The larva spins a fine silk lining to the tube and also a silken net within it. Rhythmic undulations of the body draw a current of water through the tube, carrying with it oxygen and fine particles of food in the form of protozoa and phytoplankton. The net that filters this food is periodically consumed and reconstructed. At low oxygen levels the haemoglobin in the blood makes activity possible, but in completely anaerobic conditions activity ceases altogether. The amount of oxygen present in the surrounding water has a direct effect upon the length of time spent by the larva in feeding, resting, and irrigating itself with water. There is no doubt that the presence of haemoglobin assists the chironomid larvae to remain

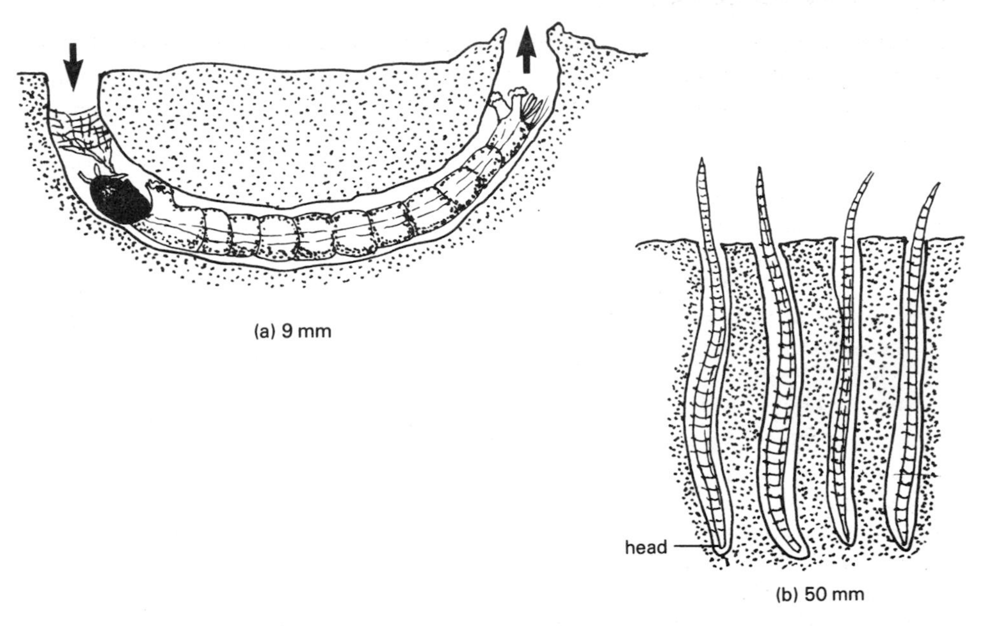

Fig. 11.17 Inhabitants of mud and silt. (a) Larva of one of the midges, *Chironomus* sp., in its silk-lined burrow in the mud. (b) Sludge worm, *Tubifex* sp. A small annelid which, like *Chironomus*, possesses haemoglobin in its blood. Several worms in their mud burrows with tails projecting

for a greater length of time within their tubes, hence reducing the risk from predators, before they have to emerge and feed on debris on the surface mud.

Pupation of the chironomid larvae takes place within the tubes and after a few days they wriggle free and float to the surface, assisted by gas bubbles which develop beneath the pupal skin. Here the winged adults immediately emerge, mate, and lay their jelly mass of eggs to start the cycle over again.

Several species of annelid worms also possess haemoglobin. The sludge worm, *Tubifex*, is an example, living head downwards in a tube constructed of mud (Fig. 11.17 (b)). Its chief predator is the larva of the fly *Dicranota* (Fig. 11.6). *Tubifex* is often found in large numbers in the mud of farm ponds enriched by the run-off from manure heaps. The tail of the worm projects from the end of the tube and waves about. It is used as a kind of gill and the greater the oxygen lack the further out the tail projects. In this way, it exposes a greater length of body in which the blood, containing haemoglobin, is circulating.

Fieldwork

1 By one of the methods described at the end of Chapter 2, measure the oxygen concentration of water at different sites in a pond (a) 5 cm below the surface (b) just above the bottom (c) in the mud. Make a list of the animals at each depth at the different sites. How does their method of respiration relate to the different levels of oxygen?
2 Examine the respiratory requirements of the water bug *Corixa*. Draw a graph of the average time 10 *Corixa* can remain below the surface without returning for air. How does this graph compare with similar recordings for the great diving beetle, *Dytiscus marginalis*? Can you account for the differences either in terms of size or method of air storage?

3 Take samples of mud from various areas in the bottom of a pond. Sort the animals present into clean water and examine their methods of respiration. List those which are dependent on (a) atmospheric oxygen (b) oxygen from their surroundings. What special devices do each possess for the extraction of oxygen for their respiratory requirements?

References

11.1 Miall, L.C. (1922) *The Natural History of Aquatic Insects*, Macmillan
11.2 Goldspink, C.R. and Scott, D.B.C. (1971) 'Vertical migration of *Chaoborus flavicans* in a Scottish loch', *Freshwater Biol.*, **1**, 411-21
11.3 Leadley Brown, A.M. (1976) *Ecology of Fresh Water*, Heinemann
11.4 Leadley Brown, A.M. (1982) 'Two Beetles of the Somerset Levels', *Som. Trust. for Nat. Cons. An. Rep.*, 24-26

12 Interpreting the freshwater ecosystem

A community is an assemblage of plants and animals which, together with the environment in which they live, is called an ecosystem.

The ecosystem of fresh water is composed of four groups of constituents: (i) The basic elements and compounds of the environment, which include water, oxygen, carbon dioxide and various mineral ions. (ii) Autotrophic plants capable of manufacturing their own food by using the light energy of the sun in their photosynthetic activities. Most of this autotrophic production occurs near the surface of a body of fresh water where light energy is available. (iii) The consumers, or the animals, both microscopic and large, which feed on plants, on other animals or on organic matter. These are the heterotrophic organisms, capable of breaking down the complex material synthesized by the autotrophs. (iv) The decomposers or saprophytes (the bacteria and fungi) which break down complex compounds found in dead and decaying organisms. They absorb some of the products of decomposition and release simple substances, such as the nutrients required by the producers.

All these processes involve complex feeding interrelationships as well as the flow of energy and other resources.

Trophic relationships

Chapter 1 described, in outline, the part played by producers and consumers in the feeding relationships occurring in an ecosystem.

Figure 12.1 is a stylized diagram to show the trophic relationships which can exist in a pond. Dissolved nutrients enter by drainage and seepage from the surrounding land and become incorporated into organic substances by the primary producers. These are the autotrophic bacteria, phytoplankton and pondweeds. The phytoplankton are consumed by zooplankton, comprising rotifers, copepods and some protozoa. Besides zooplankton, larger browsing animals such as caddis larvae feed on algae attached to larger pondweeds. Then there are the burrowers such as the bivalves *Pisidium* spp., oligochaete worms and midge larvae which feed on the ooze. All these are primary consumers. In turn, the primary consumers are preyed upon by benthic animals such as dragonfly nymphs and by plankton predators such as the phantom midge larva, *Chaoborus*. These are the secondary consumers, many of which, like *Chaoborus*, are insects that will leave the aquatic community as adults and become aerial. Tertiary consumers such as fish and water beetles are the predators of those in the lower trophic levels. Finally, the consumers and the producers, if they are not eaten by others, die and their bodies decay and contribute to the ooze, initiating the cycle once again. Indeed, at each trophic level there is bacterial breakdown of dead matter which is removed from that level and is not, therefore, available as food for the next level.

Energy flow

The source of all energy entering an ecosystem is that which is radiated by the sun. The photosynthetic activities of plants ensure that carbon dioxide is combined with water and light energy which they store as carbohydrate.

Energy passes from one trophic level to another and is eventually lost in the form of heat. Herbivores oxidize the carbohydrate of plant tissues and release carbon dioxide, water and

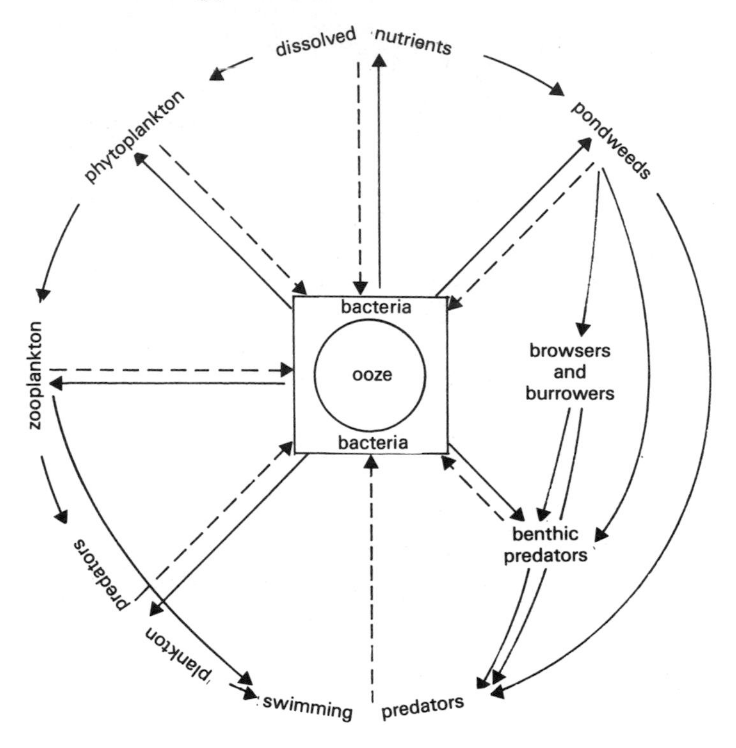

Fig. 12.1 Schematic idea of the trophic relationships existing in a pond

energy as heat. Of the ingested food, part only is transformed to actual body tissue, part is metabolized with the release of energy. The rest is got rid of as faeces. Carnivores feeding on herbivores metabolize some of the food to form body tissue and again energy is released as heat and waste material voided as faeces.

Although energy is lost to the ecosystem at each trophic level, some is stored and made available to organisms at other levels. Thus there is a transfer of energy from one trophic level to another.

Ecological pyramids

Linked with the concept of trophic levels is the important idea of the pyramid of numbers.

In terms of numbers the primary producers in any food chain are the most numerous while the primary consumers, secondary consumers and so on become progressively fewer in numbers and usually larger in size at each level.

Elton [12.1] described this phenomenon by saying 'Animals at the base of a food chain are relatively abundant, while those at the end are relatively few in number, and there is a progressive decrease between the two extremes'. This he called the **pyramid of numbers**. The classic diagram illustrating this pyramid is given in Fig. 12.2 (a). Such a state of affairs exists in ponds where there is an abundance of green algae which, so long as there is an abundance also of nutrient salts, will continue to maintain their density. But the situation

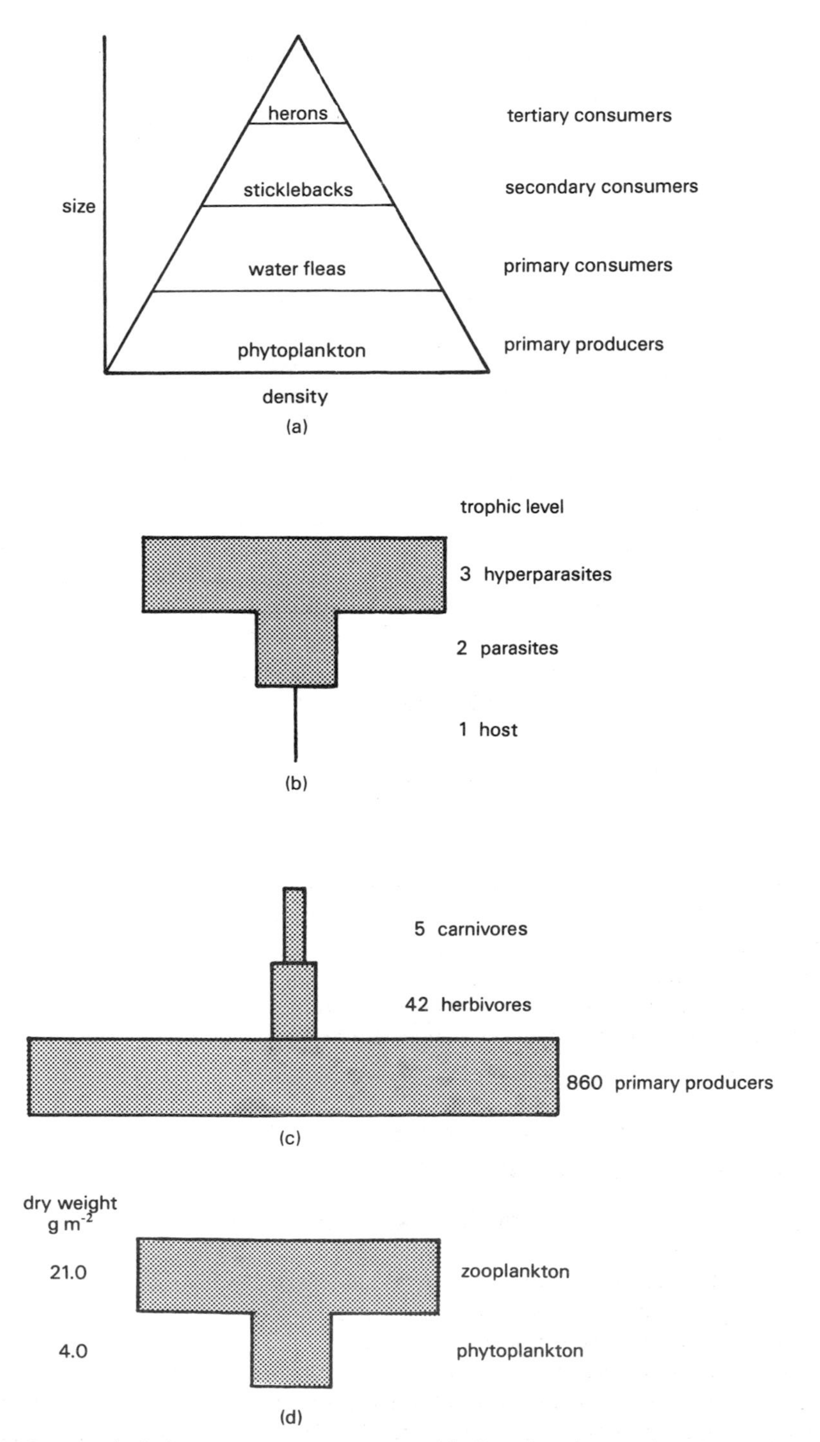

Fig. 12.2 Some ecological pyramids. (a) Possible pyramid of numbers in a pond. (b) Inverted pyramid in a parasitic chain. (c) Mean biomass pyramid expressed as g wet weight/m^{-2}, for Malham Tarn (after Philipson, 1968). (d) Inverted pyramid of biomass for plankton in the English Channel (from Odum, 1959)

in other freshwater habitats can be very different. In a stream or river, for instance, any phytoplankton able to exist and multiply in still areas near the source, will quickly be washed away by the current. The numbers in other trophic levels will therefore be affected.

In the case of parasites feeding on a host species, a pyramid of numbers becomes inverted, for a single host, such as a fish, can be parasitized by a number of fish lice. The pyramid in Fig. 12.2 (b) includes hyperparasites dependent upon parasites for their food, but in fresh water hyperparasites are uncommon.

By examining food relationships it may be possible to calculate the number of producers which support a given number of consumers which, in turn, support a given number of secondary consumers and so on. But by using the number of organisms concerned as the criteria for comparing two ecosystems, we soon find ourselves in trouble. For instance, difficulties would arise if we tried to compare the number of caddis larvae feeding on pondweed in an aquarium with the number of cows feeding on swedes in a field. Although this is somewhat far-fetched, the principle is evident, namely that if instead of *numbers* we substitute the weight, or **biomass**, of the organisms composing the different trophic levels, we can then construct a **pyramid of biomass**.

Such a pyramid was made by Philipson [12.7] for Malham Tarn. This was done by taking samples of plant and animal material from different regions of the tarn. The material was then sorted into categories according to which trophic level it belonged. The mean biomass of the sample in each category was then calculated and expressed as g wet weight m^{-2} of tarn surface.

The pyramid of biomass (Fig. 12.2 (c)), as might be expected, shows that a much higher weight of producers supports a much lower weight of consumers. More accurate results might have been obtained if the samples had been dried and their weights expressed as g dry weight m^{-2}.

In certain circumstances pyramids of biomass can be inverted. This occurs in the relationship between the phyto- and zooplankton. Calculations of biomass made during spring and early summer show that the plant plankton can give rise to a larger weight of animal plankton (Fig. 12.2 (d)). At certain seasons of the year the ratio of dry weight of zooplankton to phytoplankton may be as much as 5:1.

Bacteria may divide several times in one hour, algae several times in a day, while macrophytic pondweeds usually pass through but one generation in a year. Many crops of phytoplankton, synthesized from the original supply of nutrients, are eaten, decomposed and resynthesized several times during the year. It is therefore hardly fair to compare the total annual production, in terms of biomass, of the phytoplankton with that of the pondweeds.

Pyramids of numbers and biomass can only show the amount of material present at any particular time. This is the **standing crop** and does not tell us the *total* production of material nor the rate at which it is produced.

If instead of numbers and biomass a **pyramid of energy** is constructed, such as that in Fig. 12.3, we are using energy units as a means of expressing productivity. This provides a means not only of showing the productivity of individual organisms but of all organisms within an ecosystem, and also of comparing the productivity of different ecosystems.

The energetics of production and consumption

The amount of light energy entering an ecosystem during a given period of time is expressed as kJ m^{-2} per year. Of this, about 50 per cent is lost by radiation, the remainder being transformed into carbohydrates, fats and proteins. About 30 per cent of this energy is lost by

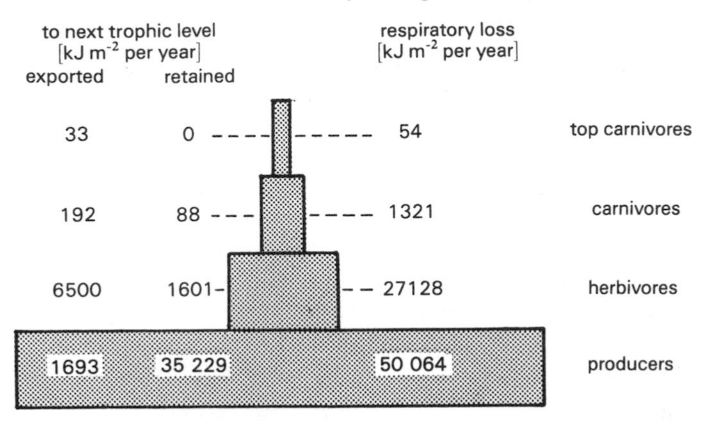

Fig. 12.3 Pyramid of energy for Silver Springs, Florida, USA (based on Odum, 1957)

the plants themselves in respiration. It is, therefore, important to know how much energy is used by an organism for its life processes and how much is lost.

In terms of the whole ecosystem the amount of light actually fixed by plants within a given time is the **gross production** (*GP*). Some of this energy is used by the plants for their metabolic processes (*M*). What is left is available for growth and is the **net production** (*NP*). So that $GP = M + NP$. But what proportion of the net production is available for the production of new tissues (T) to increase the biomass and how much of the net production is actually consumed by herbivores (C) and how much lost by death (D)? This can be expressed thus:

$$NP = T + C + D$$

The one-way flow of energy through all ecosystems results from the operation of the two **laws of thermodynamics**. These state that: (i) energy may be transformed from one type (e.g. light) into another (e.g. the potential energy of food) but can never be created nor destroyed. (ii) No transformation of energy (e.g. light to food) is one hundred per cent efficient because at each stage there is a loss of energy in the form of heat. Figure 12.4 is a simplified energy flow diagram applicable to any ecosystem. The boxes represent the population-mass (biomass) and the pipes show the flow of energy between the living units. Only about 50 per cent of the energy entering as sunlight is absorbed by the plants (the producers) and only about 1 to 5 per cent of this absorbed energy is converted by the producers into food energy. The total assimilation rate of producers in an ecosystem is called **primary productivity** (*PG* or *A* in Fig. 12.4) and is the total amount of organic matter fixed, including that used up in plant respiration during the period of measurement. **Net primary productivity** is the amount of organic matter stored in plant tissues which is in excess of respiration during the period of measurement. Net primary productivity is the amount of organic matter stored in plant tissues which is in excess of respiration during the period of measurement, and net production represents the food potentially available to the consumers. However, under favourable light and temperature conditions, when plants are growing rapidly, respiration may account for as little as 10 per cent of gross production.

At each transfer of energy from one trophic level to another or from one organism to another, a large percentage of the energy is degraded into heat. The ratio of food stored by an organism in the form of new tissue to that which is assimilated, is a measure of the efficiency of that organism. In the water flea, *Daphnia*, for instance, this has been calculated to be 50 per cent, a higher figure than that for most animals.

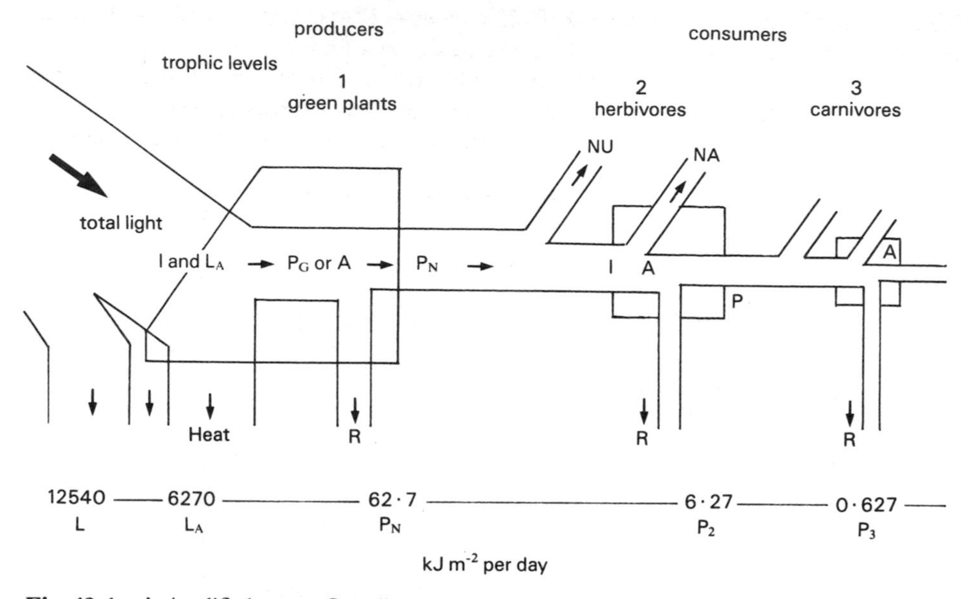

Fig. 12.4 A simplified energy flow diagram. The boxes represent the standing crop of organisms (1=producers or autotrophs; 2=primary consumers or herbivores; 3=secondary consumers or carnivores) and the pipes represent the flow of energy through the biotic community. L=total light; L_A=absorbed light; P_G=gross production; P_N=net production; I=energy intake; A=assimilated energy; NA=non-assimilated energy; NU=unused energy (stored and exported); R=respiratory energy loss. The chain of figures along the lower margin of the diagram indicates the order of incident light m^{-2} per day (after Odum, E. P. 1966)

Ecological niches

No two species have precisely the same requirements although each plays a part in maintaining the structure of the community. Every individual organism requires space in which to live and reproduce, and access to the resources of the environment. They also need food, shelter and a place in which to conceal themselves against possible enemies. Plants require light, nutrients and water. The way in which an organism exploits the environment in order to satisfy its needs is called its **niche**, which essentially is its economic status within a community. Success in attaining the necessities of life will depend on the ability of the organism to respond in the most favourable way to the environment. This will include not only modification of structure but also of behaviour.

Predator-prey relationships

The right response to environmental stimuli may well be a matter of life or death and this axiom can apply not only to individuals but to whole populations. Introduction of an organism's natural predator to a community requires the adoption of the best methods of defence. Animals and plants possessing the ability to use these methods will survive, while those which do not will perish. Avoidance of a predator is one way and makes use of any shelter afforded by the environment. The pike is often quoted as one of the top predators in a pond. Its prey are smaller fish and, providing the pond has plenty of weeds, avoidance of the pike is possible. In winter when the weeds die down, the pike will be at an advantage and may destroy large numbers of fish, thereby reducing its food supply to a point where its own life is in jeopardy. Similarly, sticklebacks feeding on water fleas in a small pond may decimate their food supply and their own numbers will decline as a result.

The food of an animal often varies with its life stages. For this reason the balance is constantly fluctuating between the numbers of predators and their prey. An extreme example of this kind of fluctuation can occur in small bodies of water such as water butts where the

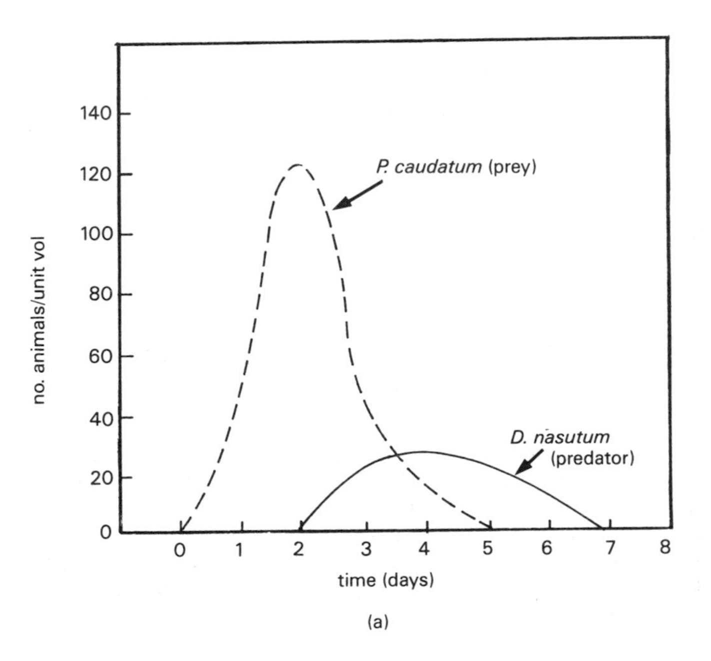

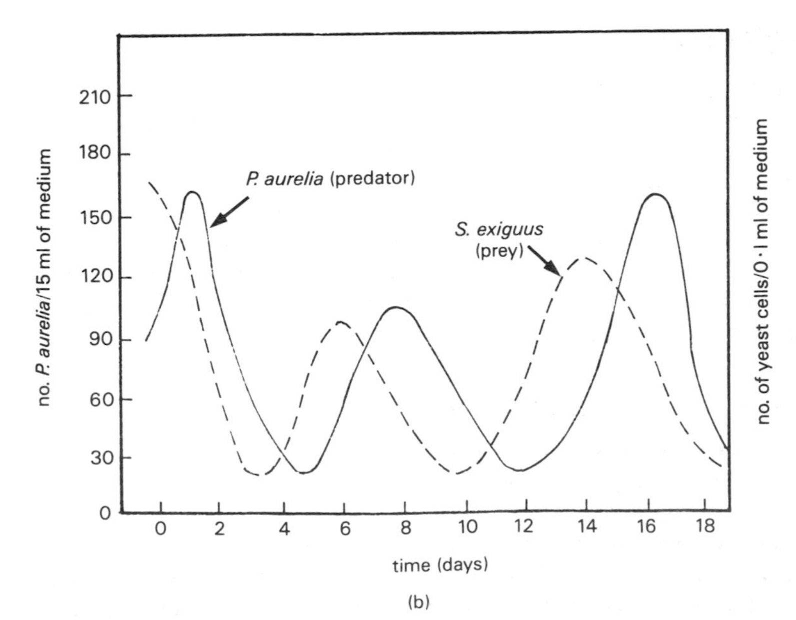

Fig. 12.5 Predator-prey relationships demonstrated with simple organisms: *Paramecium* spp., the ciliophoran, *Didium nasutum*, and the yeast, *Saccharomyces exiguus* (redrawn after Gause in Lewis and Taylor, 1967)

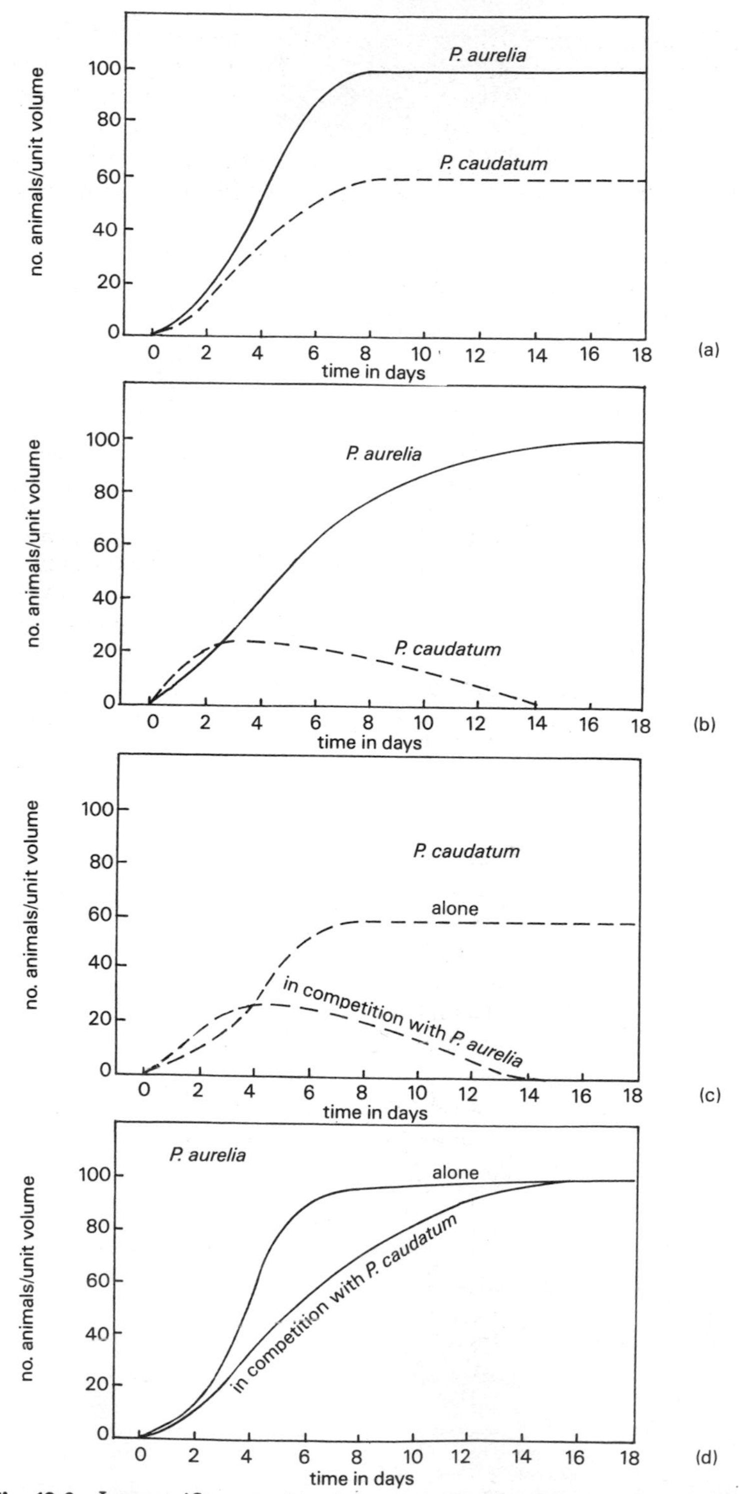

Fig. 12.6 Interspecific competition between two species of *Paramecium, P. aurelia* and *P. caudatum* (redrawn after Gause in Lewis and Taylor, 1967)

first colonists are usually algae. Rotifers arriving as eggs, possibly blown in by the wind, rapidly multiply, feeding upon the plentiful supply of algae. The algal population then becomes so reduced that it produces insufficient oxygen, by photosynthetic activities, to support the rotifers. This will result in the virtual extinction of the rotifer population and the cycle will be repeated. In a chain consisting of more links, removal of one, due to the extinction of a species, may alter the balance not only at the level at which the species became extinct but at all levels.

In all communities there is a delicate balance between predator and prey, resulting in fluctuations in populations. However, unless there is interference by man or a natural catastrophe such as flooding or drought, these fluctuations usually occur within fairly narrow limits. In the event of food shortage, organisms with a catholic diet will seek alternative food and the balance will be maintained. A clear case of the control of the balance in organisms with a restricted diet is that of a parasite and its host species, for it is not to the advantage of the parasite to destroy completely its host species.

The density of a predatory species tends to be greatest after the point of maximum density reached by the prey. In Fig. 12.5 (a) the prey population (*Paramecium*) grows until the predatory population of a ciliophoran, *Didium nasutum*, starts to multiply. The predatory population continues to increase and eventually exterminates the prey, but then has no food and so itself declines to extinction. More often, as the prey population decreases, the predatory population also decreases down to a level where the prey recovers and begins to increase. With renewed resources, the predatory numbers increase. These fluctuations can be repeated, out of phase with one another, for many cycles. This is nicely illustrated in the case of another species of *Paramecium*, *P. aurelia*, which in this case is the predator, feeding on a yeast, *Saccharomyces exiguus* (Fig. 12.5 (b)).

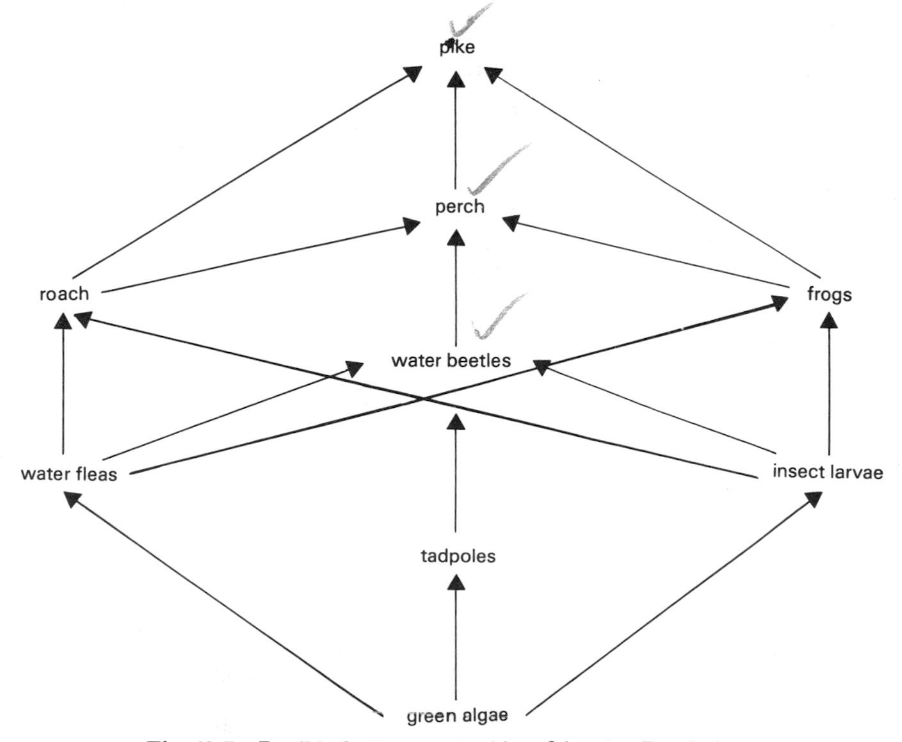

Fig. 12.7 Possible feeding relationships of the pike, *Esox lucius*

Two species living in the same environment may compete not only for food but also for living space. They may also be lethal to each other. This is illustrated by the growth of populations of two species, *Paramecium aurelia* and *P. caudatum*. Reared separately, the numbers of *P. aurelia* increase more rapidly and the population becomes stable at a density of 100 per unit volume compared with the slower growing *P. caudatum*. Figures 12.6 (a) and (b) show that when both species are present in a mixed culture, all *P. caudatum* eventually die, not because *P. aurelia* interferes with them directly, but because they eat the same food more rapidly. Figures 12.6 (c) and (d) explain the effects of competition on each species separately.

Food chains and food webs

As we saw on p. 2, in every community there exist a number of predator-prey relationships which may result in a simple **food chain**

Odum [12.4] defined a food chain as 'The transfer of food energy from the source in plants through a series of organisms with repeated eating and being eaten'.

When side links are added to a simple food chain, such as that given on p. 2, it quickly

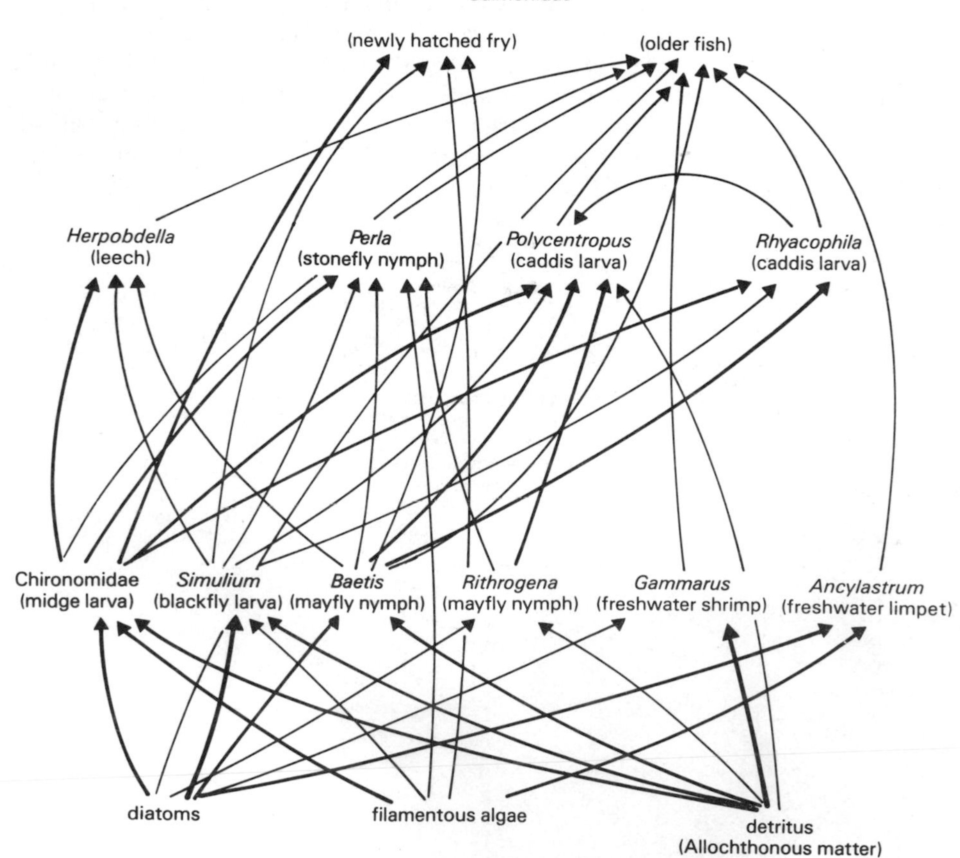

Fig. 12.8 Relatively complex food web of trout and other salmonid fish in a stream community. An important primary source of food is detritus derived from allochthonous material. (Based on data tabulated for streams in the Dee River system, Wales, by R.M. Badcock in J. Anim. Ecol, **18**: 193–208, 1949. From Russell-Hunter, 1970)

becomes a **food web** (Fig. 12.7). Food webs in fresh water can be more complex, as in the examples given in Figs. 12.8 and 12.9. An increase in complexity of a food web such as these, probably creates increased stability of the community. Introduction of top carnivores, such as the otter or man at the top of a web, would add further trophic levels.

Tracing food chains

It may be possible within the confines of an aquarium to make direct observations of predator–prey relationships. Perhaps it is possible to note that a caddis larva is browsing on algae and that a stickleback eats the caddis. However, this is a very inadequate analysis.

A more exact estimation of feeding preferences can be obtained by the careful analysis of gut contents. This involves the collection and dissection of a large number of individuals of each species, a procedure not to be commended. In any case, such an examination may only

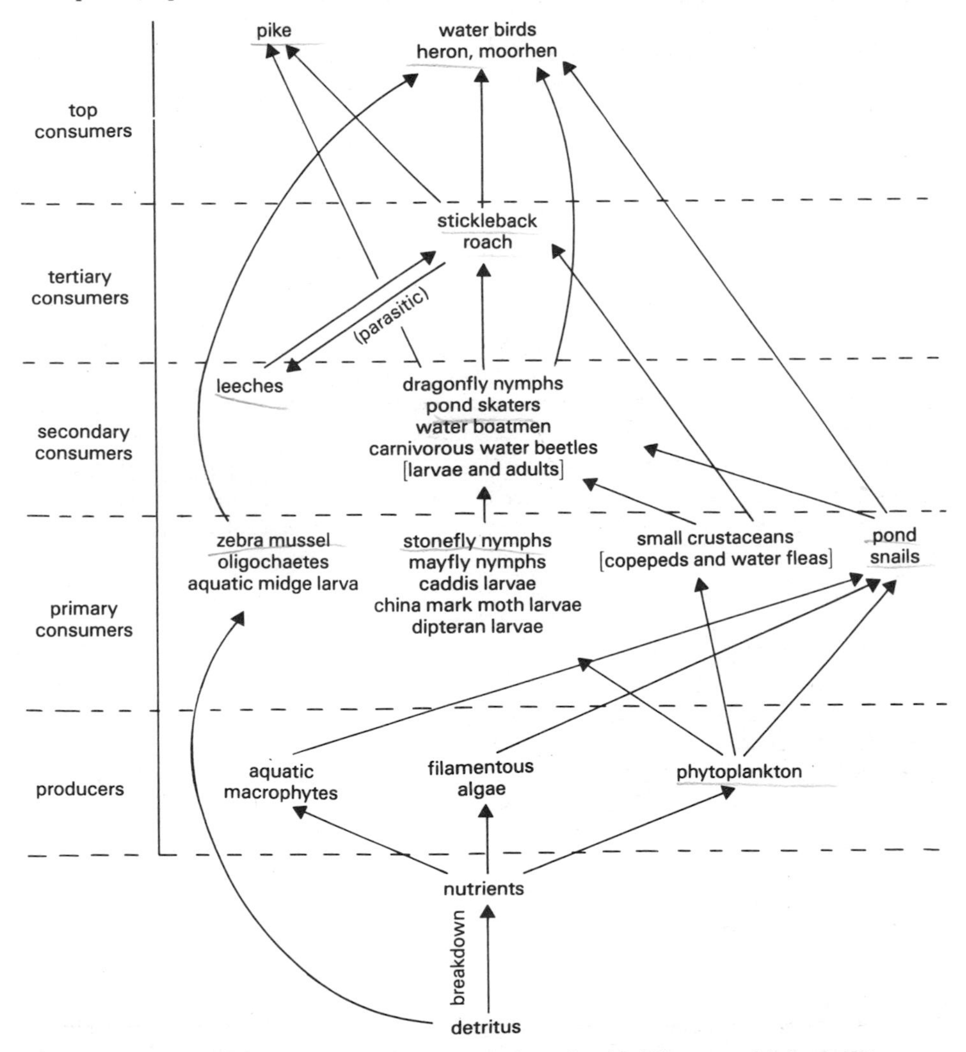

Fig. 12.9 Part of a general food web in a pond to show also the different trophic levels. The excrement and dead bodies of organisms contribute to the detritus as well as any allochthonous material falling into the pond

make possible the identification of those organisms with hard skeletal parts. Those with only soft bodies may escape notice altogether and plant juices, which are the diet of many herbivores, are rapidly assimilated after ingestion.

The **precipitin test** has been used with some degree of success in tracing the food relationships of certain freshwater species. (See Reynoldson and Young [12.10]; Young, Morris and Reynoldson [12.15]). This method was used in tracing the diet of the flatworm, *Dendrocoelum lacteum*. Gut analysis of this species revealed undigested setae which were identified as belonging to oligochaete worms. Direct observation of the feeding habits of *D. lacteum* showed, however, that it preferred the water louse, *Asellus*, ingesting only the soft tissues and body fluids. The precipitin test was applied which indicated that the food of *D. lacteum* consisted largely of *Asellus*. This result would not have been revealed by gut analysis alone.

Radioactive isotopes have been used in tracing terrestrial food chains. Plant foliage of a single species is 'labelled' by spraying onto it phosphorus-32. (^{32}P). Animals living on the plants are then tested for the presence of ^{32}P. Any showing radioactivity will then have been either directly or indirectly dependent on the plants as the original source of food. Any radioactive substance which can be used to 'label' a primary food source, and which can be subsequently traced through other organisms, can be of value in deciphering the stages of a food chain and might possibly be used in this manner for tracing aquatic food chains.

The ecosystem in action

In complex ecosystems a complete picture of energy flow is only possible if data recording population dynamics, life cycles, food assimilation and respiration of the community is available. Obtaining such statistics may take many years and assumes a reasonable degree of stability within the ecosystem.

The rate of production is important, as well as an understanding of the effect of an alteration of various parameters within the ecosystem. For example, in an enclosed, artificial trout pond, rapid breeding will cause a population build-up to a level where there is just enough food to maintain the metabolism of the fish but not enough for further growth. Net production in this case would be low but the standing crop high. To increase productivity there must be more food made available to increase energy input added either directly or indirectly in the form of nutrients. The more obvious remedy would be the removal of some of the stock to allow the remainder additional food.

The production of a complete energy budget for an ecosystem is a formidable task which would be virtually impossible for large complex ecosystems. Lindemann [12.3] was one of the first workers to attempt such measurements and it was because of his work that others were encouraged to investigate the dynamics of other ecosystems, improving on Lindemann's methods.

Teal [12.14] made his classic study of a small spring, called Root Spring in Massachusetts which was about 2 m in diameter and only 10-20 cm deep. Being small and containing a relatively simple community of organisms, the spring was virtually a 'laboratory ecosystem'. Another important factor was that there was a thick deposit of mud on the bottom covered with a layer of dead leaves in which most of the animals lived. Chief among these were detritus feeders such as *Limnodrilus* sp., *Asellus* sp., *Pisidium* sp., *Physa* sp., *Crangonyx* sp., and *Calopsectra* sp.

Since most of the plant material forming the food of these herbivores was allochthonous in origin, an estimate of the energy entering the ecosystem was made by measuring the energy produced by the breakdown of the debris as well as the gross production by the plants living in the spring, which were diatoms, filamentous algae and duckweed.

Traps, in the form of wooden trays, were constructed to catch the allochthonous matter and the calorific value of this was calculated to be 9823 kJ m^{-2} per year. The gross production of energy by the plants, which were all either microscopic or very small, was measured by the dark and light bottle method (Appendix 3).

Open glass cylinders, 17 cm in diameter, were pushed into the mud at the bottom of the spring so that their tops projected 2–4 cm above the surface of the water. They were then filled with spring water and covered with an airtight seal. At the beginning and end of every 24 h period during the investigation, the oxygen content of the water in the cylinders was measured using the Standard Winkler technique (Appendix 2B). At the conclusion of the experimental period the cylinders were enclosed in black boxes to arrest photosynthesis. The oxygen content was then again measured at the beginning and end of a set period. From the results obtained with the light and dark cylinders, the amount of oxygen evolved by photosynthesis was calculated and from this a figure for the gross production was derived and expressed as kJ m^{-2} per year. Teal's results were as follows:

	kJ m^{-2} per year
Gross productivity of plants	2968
Estimated respiration of plants	230
Wet productivity of plants	2738
Energy produced by allochthonous matter	9823
Energy not used by herbivores	2951
Energy utilized by herbivores	9614

Teal went on to investigate how the energy taken in by the herbivores was used. The problem also involved the calculation of how much of this energy, if any, was used to increase the standing crop of herbivores and also how much was passed on to the carnivores and whether the remainder was dissipated as heat during respiration.

To determine these figures it was necessary to ascertain the numbers, biomass, respiratory rate and mortality for each of the principal animal species and, in the case of insects, their moulting losses. Monthly population estimates were obtained by taking random samples of the mud and identifying the chief organisms which were hand sorted, counted and weighed. Their calorific values were then determined by a chemical method. More modern miniature bomb-calorimeters are able to cope with samples as small as 5 mg dry weight [12.8]. Special traps were used to catch emerging insects and their biomass calculated per square metre.

Table 12.1 shows the population sizes per square metre for the midge, *Calopsectra dives*, one of the herbivores present in Root Spring. This was a record of one species of insect with larvae resident in the mud debris of the spring whose adults emerged and left the enclosed

Table 12.1 Population of larval and emerging adults of *Calopsectra dives* in Root Spring, Concord, Mass. in 1965 (after Teal, 1957)

	Jan	*Feb*	*Mar*	*Apr*	*May*	*June*	*July*	*Aug*	*Sept*	*Oct*	*Nov*
Larvae											
Number m^{-2}	0	0	0	0	1700	89 500	65 000	57 000	200	0	0
kJ m^{-2}	0	0	0	0	8·78	168·87	237·42	366·17	0·42	0	0
Adults											
Number m^{-2}	0	0	0	0	13	170	953	3464	13 250	533	
kJ m^{-2}	0	0	0	0	0·.08	1·03	5·77	20·90	80·67	3·24	

ecosystem of the spring. The table shows that during the period January to November, the larval standing crop grew from zero in April to its maximum in August of 336.17 kJ m^{-2} falling to zero in October. This shows that there was no energy change over the year and, more importantly, that none of the energy consumed by *Calopsectra* was used to increase the standing crop of the species. There remained calculations to be made of the amount of energy contained in the larval skins and that of any larvae which died between one sampling occasion and another, as well as the amount of energy dissipated as heat of respiration. All these measurements were needed before an energy balance sheet could be constructed for *Calopsectra*. Similar procedures were adopted for all principal faunal species in the spring, from which it was possible to construct an energy flow diagram for the whole ecosystem of the spring (Fig. 12.10).

From the results it is important to note that of the 9614 kJ m^{-2} per year consumed by the herbivores, only 2738 kJ were produced within the ecosystem, the rest entered as debris; from outside. Of the 12866 kJ m^{-2} per year entering the spring, 9133 kJ was dissipated as heat and 138 kJ lost via emerging insects. The rest was deposited in sediments, much of which would eventually be washed away.

Although the study of Root Spring involved a vast amount of work, the ecosystem, by comparison with others, was relatively simple. Odum [12.6] worked with a team on a spring in Florida, Silver Springs, which was another small ecosystem. His objective was to study the trophic structure of the spring and its productivity. Odum constructed an energy budget by measuring the total energy entering the spring, the energy passed on by one trophic level to another and the heat lost at each level. From his results he constructed an energy flow model (Fig. 12.11). Comparing this with Fig. 12.10 for Root Spring it is evident that in Root Spring most of the energy entered the spring from the outside as allochthonous debris, whereas in Silver Springs most of the heterotrophic food was produced within the spring by the autotrophic plants. In other words some heterotrophs fed upon living plants while others

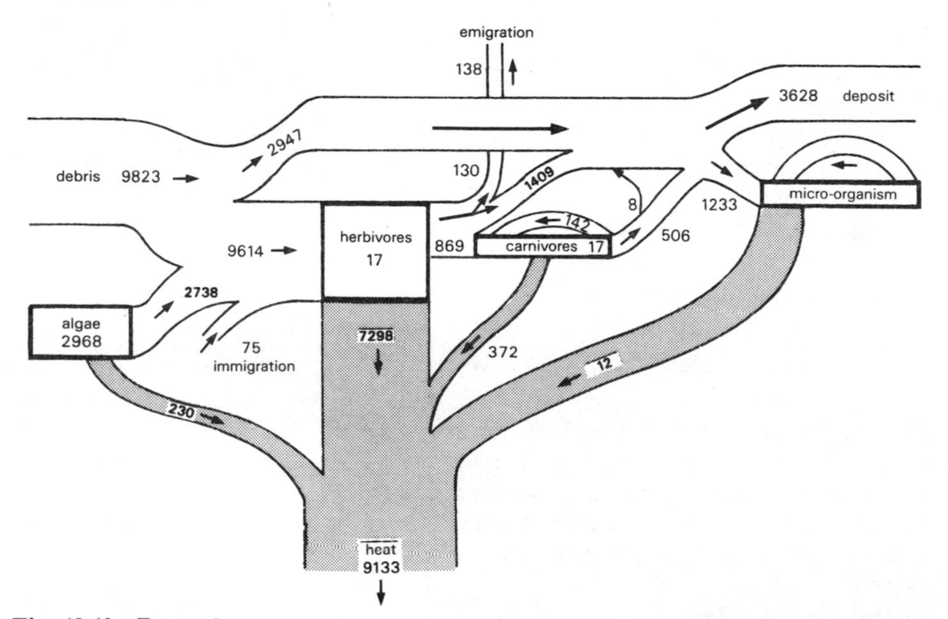

Fig. 12.10 Energy flow diagram for Root Spring, Concord, Mass., in 1953–4. Figures in KJ m^{-2} per year. Note that the figures relating to the boxes do not represent standing crops. The figure for algae represents gross production over the year while the figures for carnivores and herbivores show the net change in standing crop (adapted from Teal, 1957)

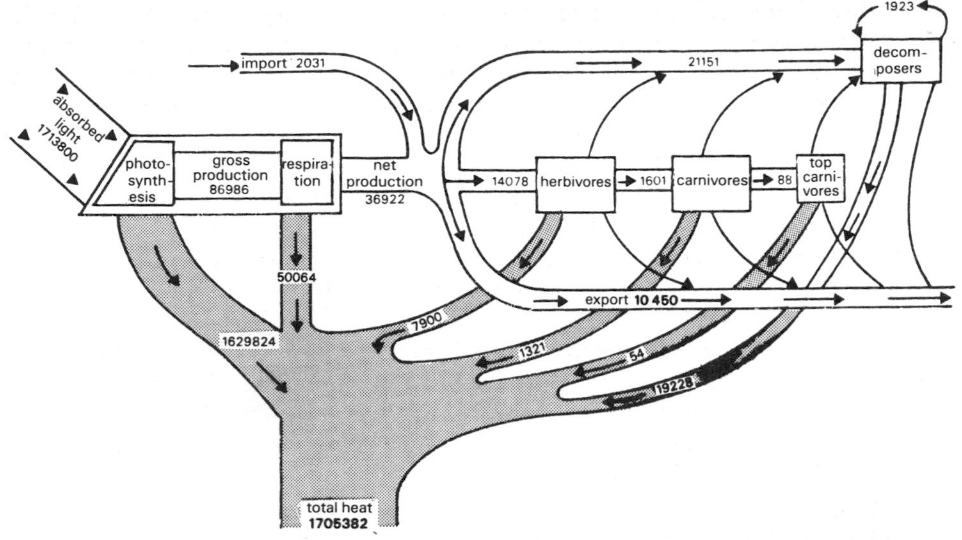

Fig. 12.11 Energy flow diagram for Silver Springs, Florida. Figures in KJ m^{-2} per year (adapted from Odum, H. T. 1957)

fed on dead plant material. There were, therefore, two basic food chains operating in each ecosystem: a **grazing food chain** (p. 2), which consisted of herbivores feeding on living plants, and a **detritus food chain** (p. 2), consisting of animals feeding on dead plant material. It is also true to say that the dead bodies of animals, which were once part of the grazing food chain, eventually became incorporated in the detritus chain along with their faeces. The evidence available suggests that of these two feeding chains, that of the detritivores (decomposers) was of the greater importance. This is certainly borne out by the results obtained from Silver Springs, where the flow of energy through the grazing chain was calculated to be 14078 kJ m^{-2} per year, while through the detritus chain it was 21151 kJ m^{-2} per year.

More ambitious studies have been carried out recently, involving a number of freshwater ecosystems in different parts of the world. These have necessitated improved methods of measurement and have revealed greater insight into the processes involved in the transfer of energy within aquatic ecosystems.

The decomposers

The organisms which bring about decay are the bacteria and fungi which operate upon dead and decaying plant and animal matter. Great importance, as we have seen (p. 10), must be attached to the role of these decomposers and the results of their work, for this represents a large proportion of degraded material within any ecosystem. The nutrients released as the result of decomposer activity are made available to plants, the primary producers, for incorporation within their tissues. Detritus derived from allochthonous material falling into the stream from overhanging vegetation is the primary food source in the food web of the Salmonidae shown in Fig. 12.8 on p. 136.

Ecological efficiency

The question now remains of how efficiently energy is transferred from one trophic level to another. In an ecosystem the amount of energy made available to a predator at the next trophic level divided by the amount of food ingested by the prey is a measure of the **ecological efficiency** of those levels of the ecosystem.

Philipson [12.7], in his survey of Malham Tarn, found that the ratio of primary consumers to primary producers was 1:20 and that of the secondary consumers to the primary consumers was 1:8 (Fig. 12.2 (c)), both being derived only from measurements of standing crop biomass. Since herbivores have a less efficient digestive system than that of carnivores, one would expect the former of the two ratios to reflect this greater inefficiency.

Few measurements have so far been made to determine the ecological efficiencies of natural systems and figures range from 5 to 30 per cent, but Slobodkin [12.12] considers 10 per cent to be fairly consistent among a range of conditions investigated.

Theoretically, in a perfect ecosystem, there should be sufficient organisms at each trophic level to consume, during the year, the organisms of the level below, down to a point where they are never numerous enough to run short of food nor depleted in numbers to a point where any of their food goes to waste.

References

12.1 Elton, C. (1927) *Animal ecology*, Sidgwick and Jackson
12.2 Lewis, T.Y. and Taylor, L. (1967) *Introduction to Experimental Ecology*, Academic Press
12.3 Lindemann, R.L. (1942) 'The trophic-dynamic aspect of ecology', *Ecology*, **23**, 399-418
12.4 Odum, E.P. (1971) *Fundamentals of Ecology*, 3rd ed., W.B. Saunders Co.
12.5 Odum, E.P. (1966) *Ecology*, Holt, Rinehart and Winston
12.6 Odum, H.T. (1957) 'Trophic structure and productivity of Silver Springs, Florida', *Ecol. Monogr.* **27**, 55-112
12.7 Philipson, G.N. (1968) 'Ecological pyramids: A field study at Malham Tarn', *School Sci. Rev.*, **50**, No. 171, 262-78
12.8 Phillipson, J. (1964) *Oikos*, **15**, 130-39
12.9 Phillipson, J. (1966) *Ecological Energetics*, Studies in Biology No. 1, Arnold
12.10 Reynoldson, T. B. and Young, J. O. (1963) *J. Animal Ecol.*, **32**, 175-91
12.11 Russell-Hunter, W. (1970) *Aquatic Productivity*, Collier-Macmillan
12.12 Slobodkin, L.B. (1960) 'Ecological relationships at the population level', *Amer. Nat.*, **94**, 213-36
12.13 Slobodkin, L.B. and Richman, S. (1960) *Ecology*, **41**, 784-85
12.14 Teal, J.M. (1957) 'Community metabolism in a temperate cold spring', *Ecol. Monogr.* **27**, 283-302
12.15 Young, J.C. Morris, I.G. and Reynoldson, T.B. (1964) *Arch. Hydrobiol.*, **60**, 366-73

13 Man and freshwater communities

Of all habitats, fresh waters are probably the most vulnerable and most subject to change by man's activities. The still waters of lakes and ponds are the receivers of human waste in all its forms, while rivers provide the means of removal and distribution of such waste to the sea.

As with other habitats, fresh water is a complex of chemical, physical, and biotic processes, tending, when faced with a change in external conditions, to adjust to a state of maximum *homeostasis*, that is the maintenance of constant internal conditions counteracting fluctuations in the external environment. In the ecosystem of fresh water, this is achieved by the continuous operation of natural selection pressures upon the organisms living in that environment. Although man constantly seeks to improve, by some form of management, an ecosystem which has achieved a natural homeostasis, this is never possible without causing an imbalance. Increasing fish populations or clearing weed beds, for instance, results in a series of repercussions which must be counteracted in one way or another in order to restore the balance. Again, change in use of a flooded field area by drainage drastically alters the nature of the land and may involve building up river banks to contain the flood water — a short term answer which may in the end prove unfavourable.

In practical terms conservation of fresh water really means that we must make the best of what is left after rivers and their catchment areas have been used to satisfy the needs of human populations.

The effect of population increase combined with higher living standards have placed ever larger demands on available water resources. In many areas of Great Britain, as well as in other countries, these resources do not meet demand. But to these inadequacies we can add other problems which arise directly or indirectly as the result of our growing population. Pollution, in all its forms, over-extraction of water from rivers, drainage, and various agricultural activities are all threats to natural water courses.

In neolithic times man altered the landscape by cultivating, burning, and felling trees, all of which inevitably had their effect. Since those times this has been repeated in many areas of the world. Cultivation leads to soil erosion and hence to silted water courses, while clearing of forests increases the rate of run-off and may therefore lead to quite dangerous fluctuations in the speed of flow of a river.

It is a sad fact that more rivers and lakes have been, and are being, degraded than created. The chief causes for this are river management by drainage and the increase of nutrient loadings of rivers and lakes resulting in eutrophication (p. 148 et seq).

Drainage can involve dredging and straightening of a river, the building of flood banks, and clearance of weed in an effort to move the river water along to the sea as fast as possible. Other threats to the ecosystem are over-fishing, over-stocking with fish, and the introduction of exotic species and of toxic pollutants, not to mention the pressures of recreational pursuits.

Sources of pollution

The administration of waterways in Britain is at present in the hands of various Water Authorities who are responsible for the rivers, reservoirs, and other bodies of water within their own areas. Originally, these authorities were primarily concerned with the chemistry of the waters under their control. To this aspect has now been added the effects of physical and chemical factors on the biology of the fresh waters. This has led most Water Authorities to employ biologists with an important part to play in assessing the effects on pollution of the living organisms present.

In a densely populated and highly industrialized country such as Great Britain, both industry and sanitation produce effluents which must flow somewhere before reaching the sea. Water usage has reached such high levels that in many places most of the water arriving in the lower reaches of a river has been used for some purpose or another. The water of the River Thames, for instance, is said to be drunk and excreted five or six times between source and mouth. Indeed there are some rivers whose major source is now a sewage works and, except in wet weather, were it not for the sewage these rivers would dry up altogether.

Purification of sewage and methods of treatment of most other kinds of polluted effluents are now applied to our waterways. Nevertheless, few can be rendered entirely unharmful unless they receive considerable dilution after treatment. This is one of the reasons why extraction of water from a river by industry can grossly alter the effects of treating harmful effluents further downstream, especially after a prolonged drought, such as we suffered in 1984. In that year abstraction of water in the River Torridge, in Devon, above a large dairy plant, caused gross pollution below the plant, intensified by drought conditions. Monitoring of the situation might have avoided this catastrophe. Now, aware of such dangers, town planners and engineers carefully consider the siting of new factories.

The causes of pollution are many, ranging from the direct entry to a waterway of untreated pollutants, an unlikely event, to seepage from catchment areas of poisons such as farm sprays, sheep dips, drainage from rubbish tips or silage pits, and slurry. The excessive use of DDT, one of the organochlorine insecticides, during the 1960s and 70s had the most damaging effect of all the insecticides sprayed on crops to control aphids and other insect pests. Manufacture of DDT is now banned although it is currently permissible to use existing stocks. Paraquat, another harmful insecticide, is still in use as a farm spray.

Effluents fall into different categories which have varied polluting effects.

Mining and quarrying, of china clay for example, produce washings which are **inert suspensions** of fine particulate matter. Several types of industry produce effluents containing sulphides, sulphites, and ferrous salts which seep underground, eventually reaching waterways. Those actively pumped into rivers produce serious pollution, for the acid ferrous salts become oxidized by bacterial action, precipitating ferric hydroxide, which becomes an inert suspension. A deep rusty-red deposit in a stream is evidence of ferric hydroxide.

Oil from factories and garages or from refuse tips can be washed into streams covering large areas of water surface.

Organic pollutants have various origins. Dairies, silage, manure pits, slaughter houses, laundries, breweries, tanneries, paper mills, and many other industries all produce their own form of organic waste. One of the most serious is domestic sewage. Most of these organic compounds are attacked by aerobic bacteria making use of the dissolved oxygen in the water to oxidize the compounds, thereby creating an oxygen deficiency.

Some industries also produce **poisons** in solution. These include chromium salts from tanneries, phenols and cyanides from gas works, insecticides from sheep dips, and many others. The heavy metals such as copper and lead are soon precipitated, but others are more persistent.

Industries, such as steel works, and electrical power stations use river water which is pumped through cooling towers. The discharges from these of **warm water** initially provide a habitat for large bacterial populations, which create a large oxygen demand (Fig. 13.1). Treatment by chlorination in carefully regulated doses avoids the presence of free chlorine in the returned water. However, chlorine readily combines with any organic matter

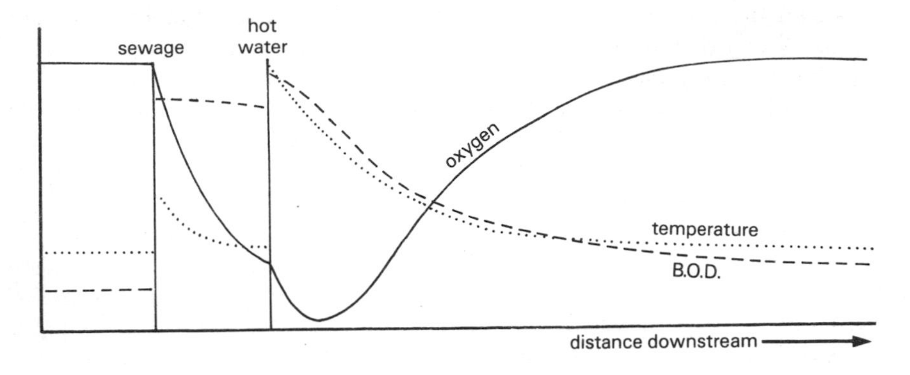

Fig. 13.1 Diagram to show the effects of a heated effluent on an organically polluted river (after Hynes, 1960)

present to form toxic compounds so that, even if no free chlorine is present, an effluent may still contain poisons.

Polluted effluents can now be treated in various ways to reduce their toxic effects but only in a few can such treatment be totally effective. The reason for this is that, as the concentration of substances in solution (other than those in suspension) falls, removal below a certain concentration becomes increasingly difficult and expensive, so that the final effluent may still contain small amounts of pollutant.

Sewage disposal

The proper disposal of sewage is of the greatest importance. The techniques used in modern sewage works are complex but rely on a few basic biological principals. Briefly, the first stage involves sedimentation of the sewage, which is then passed through a bed of clinkers. This is done by the use of perforated pipes which rotate above a clinker bed, sprinkling the liquid sewage in fine sprays onto the clinker. Oxygen, which is readily available, renders the aerobic organisms growing on the surface able to operate in reducing the organic content of the sewage by oxidation to simple salts. The organisms taking part in this process are various bacteria such as *Zoogloea*, *Sphaerotilus*, and *Beggiatoa* as well as micro-fungi and protozoan vorticellids. Feeding and growing, these micro-organisms gradually fill up the instertices of the filter bed and eventually block up the bed, reducing its efficiency by cutting off the air supply. Normally, this is prevented by the breaking away of the growths to produce suspended matter. Sludge worms (Tubificidae) and insects such as the sewage midges, *Metriocnemus* spp., feeding on the encrusting growth also help to break it up.

From this brief account it is clear that chemistry and biology are inextricably bound together in the treatment of water pollution, neither discipline being able to operate without the other.

Measuring pollution

In most cases of pollution the immediate result is a reduction in the oxygen content of the water as the result of aerobic bacterial breakdown. The degree to which oxygen levels fall can be critical to many freshwater organisms (p. 13), each species having its own range of tolerance.

Nearly a century ago, in 1898, the Royal Commission on Sewage Disposal was set up and during the succeeding fifteen years it made a number of important recommendations which were widely accepted. A large selection of rivers were investigated and results showed that the dissolved oxygen absorption tests, now known as the Biological Oxygen Demand Test (BOD), gave a fair measure of a river's degree of pollution. This test is now the standard method for measuring oxygen levels. The test depends upon the purely arbitrary measurement of oxygen taken up by a sample of effluent, or river water, during a period of 5 days at 20°C. The method involves the dilution of the sample with well-oxygenated water, in sufficient quantity to ensure that a 50 per cent oxygen saturation will remain after 5 days of incubation in a sealed bottle. The initial oxygen content is measured (a) and a portion of it is then stored at 20°C in the dark (to ensure that no algal photosynthesis takes place). After the 5-day period, the oxygen content is again determined (b) and, from the difference between (a) and (b), the BOD of the original sample can be calculated. Allowance must be made for the degree of dilution and the oxygen demand of the diluting water.

For sewage, the 5-day incubation period is insufficient since even a 10-day incubation represents only 90 per cent of the total BOD. However, the 5-day test is useful for comparing effluents.

The lengthy time taken by sewage to break down, even at 20°C, demonstrates how far downstream the effects of deoxygenation can extend (Fig. 13.2). These breakdown processes will be further slowed down in the cool water of a river, extending the range of the effects of deoxygenation. The value of BOD tests for problems of pollution are discussed in detail by Phelps [13.4], and the biological aspect of pollution is dealt with in a paper by Chandler [13.2].

The chief disadvantage of the BOD test is that it takes a minimum of 5 days, after sampling. In practice, this can be a drawback for a sewage works manager who must be able to monitor the plant at short intervals. The potassium permanganate test, therefore, is often used. This test is based on the amount of oxygen absorbed from potassium permanganate by an effluent sample. The test is run at 27°C for four hours [13.3].

In the United States it is usual for the polluting effects of various industrial effluents to be stated as 'population equivalents'. Phelps [13.4] quotes a number of examples. For instance, each individual human being contributes a total oxygen demand of 115 g per day. Table 13.1 gives other examples.

Table 13.1 Examples of total oxygen demand per day of some industrial processes (based on figures from Phelps, 1944)

Industrial process	*Equivalent no. of humans*	*Total oxygen demand* (g^{-1} per day)
Tanning one hide	18	2070
Slaughtering one cow	21	2415
Making 45 359 g butter	34	3910
Manufacturing 1016 kg strawboard	1690	194 35

The biological effects of pollution

Despite the application of various methods of pollution control, some undesirable substances remain. These are either poisons or organic matter, which may act as a poison. Both exert their effects on freshwater organisms.

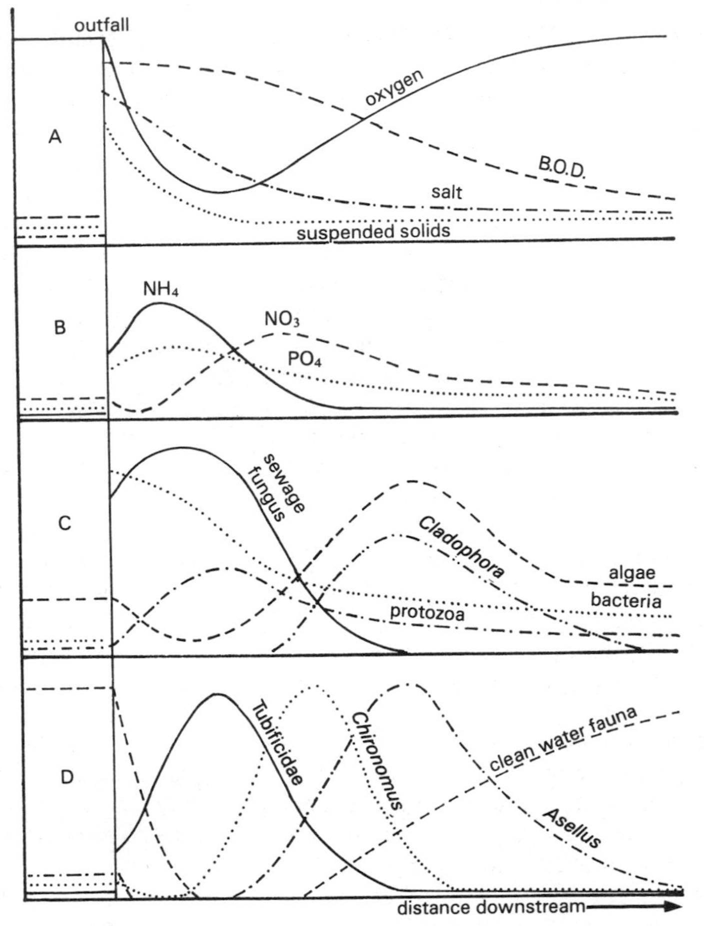

Fig. 13.2 Diagram to show the effects of an organic effluent on a river as one passes downstream. A and B chemical and physical changes, C changes in micro-organisms, D changes in macro-organisms (after Hynes, 1960)

Poisons rarely occur singly. More often there are several poisonous substances present which react to increase each other's toxicity. For instance, in effluents from gas works, chlorine acts on the thiocyanates present to produce a poison more toxic than either of them. Toxicity is also affected by such factors as temperature, oxygen content, pH, and dissolved salts. Investigation of toxic effects in a river is, therefore, complicated by the various environmental conditions.

Assessing the effects of poisons on freshwater organisms is greatly complicated for a number of reasons, the chief one being that each species reacts in a different way. More work has been done on fish and their reaction to poisons than on any other group and it has been shown by various workers that species of fish react to different toxins, and to different strengths of toxin, in a variety of ways which are not always predictable. The same applies to invertebrates which may be difficult to keep for long periods under observation in laboratory conditions. Nevertheless, figures are available for the sensitivity of some species to toxins. The highest concentration of zinc tolerated by *Limnaea pereger*, for instance, is 0.2 mg^{-1} while water boatmen, certain stoneflies, and caddises can tolerate up to 500 mg^{-1}.

Algae, like animals, are affected by poisons to different extents and in polluted waters some species reappear sooner than others. Very early on, in the investigation of pollution by various metals in Welsh streams, Carpenter [13.1] showed that in these streams, which contained ore mining wastes such as lead and zinc, almost the only plants to survive were the algae *Batrachospermum* (Fig. 8.1 (a)) and *Lemanea*. Fish and most invertebrate species were absent. As the ores became leached out and as one proceeded downstream to regions of greater dilution, a biological succession was evident. First to appear were various species of algae, worms, and flatworms. Later, when no more metals were present, starwort, water crowfoot, some fish, molluscs, and other invertebrates appeared.

Figure 13.3 shows a hypothetical case in which the concentration of a poison after entering a river declines due to dilution, precipitation, or destruction. Animal species, at first eliminated, reappear downstream in increasing numbers. Algae which survive may initially build up in numbers because of the absence of herbivores, and then decline.

Acid rain

Any account of pollution of fresh waters would not be complete without reference to the much-discussed problem of acid rain. This is caused by sulphur dioxide and nitrous oxide in the smoke and fumes emitted from power stations and factories. Both these substances readily dissolve in water droplets in the atmosphere to form sulphuric acid and nitric acid respectively. It is possible for the droplets to be carried great distances before eventually falling as rain. Lakes, rivers, and streams can become contaminated by these acids, causing losses to fish stocks. Acid rain falling on soil affects forests and crops by lowering the pH of the soil.

Measures are already being implemented to reduce these pollutants but may well prove to be insufficient in preventing the build-up of acids and the consequent effects on forests, crops and bodies of fresh water, both in this country and abroad [13.5, 13.6].

Eutrophication and nutrient balance

Of the nutrient ions needed for plant growth, the most important are nitrate and phosphate. Release of these ions can have a profound effect on plant growth in rivers and lakes, causing increased development of microscopic algae. Effects are far greater in the still waters of lakes

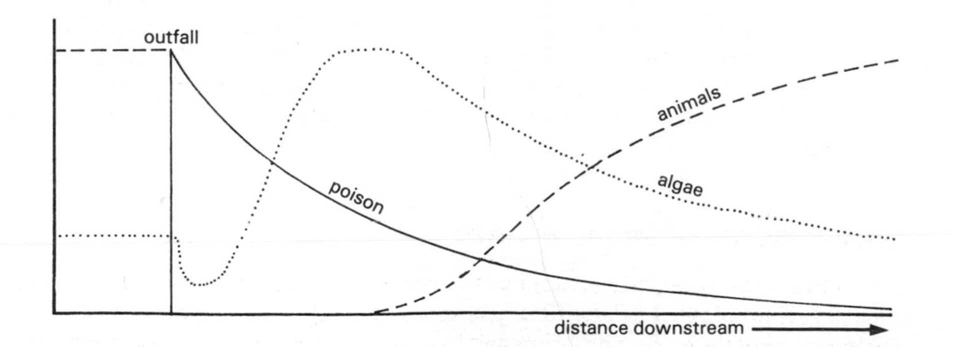

Fig. 13.3 Diagram to show the decrease in the concentration of a poison in a river and the changes in the numbers of algae and species of animals (after Hynes, 1960)

if they receive excess amounts of nutrients either directly from run-off or via a river or stream. This results in eutrophication, or the massive production of algae, which increases the turbidity of the water. Some species of blue-green algae produce substances toxic to fish and a whole train of events in the ecology of an affected lake takes place.

Increased algal growth, due to nutrient enrichment, causes a decline in submerged macrophytes. Blanketing of these larger plants by filamentous algae such as *Cladophora* (Fig. 8.1 (b)) reduces the amount of light reaching them. However, the decline of the macrophytes is not just a simple problem of shading for they secrete inhibiting substances which affect phytoplankton, restricting their growth. If nutrient loading continues to increase, growth of epiphytic algae and filamentous species continues, restricting light to the submerged macrophytes. Less energy is then available for the production of phytoplankton inhibitors, allowing them to increase once more. Eventually eutrophication results in the elimination of all macrophytes and an affected lake presents a dead appearance with a covering of green slime.

Fish farming

Recently there has been an enormous increase in the number of fish farms. These range from small trout farms, usually sited in valleys in which streams have been dammed to enclose a series of artificial ponds, to the much larger commercial enterprises involved in salmon farming.

As in other farmed species, production is estimated as the increase in weight per unit time and this must take into account losses due to disease and predators. Rate of production will depend upon the right food balance as well as upon the numbers of fish being raised. Usually the amount of food supplied far exceeds that actually consumed and wastage, combined with excretion, can cause high pollution levels. This is particularly evident in the sea lochs now being used for salmon farming up and down the west coast of Scotland. Lochs are usually selected for their proximity to suitable sources of running fresh water where the first stages of growth must take place. Some, which are relatively landlocked and therefore unexposed to westerly gales, are favoured for anchoring the mesh-enclosed rearing cages in which the salmon are fed on pelleted food. Much of this falls through the cages and accumulates on the sea bed. In shallow waters this results in over-enrichment and pollution.

It is a well-known fact that the Atlantic salmon has a strong homing instinct to return to the same river where it spent its early development. However, some farm-reared salmon are now being used to stock rivers and there is evidence that, due to genetic differences, the homing instinct of such hatchery fish has been destroyed. It was once thought that the Atlantic salmon was a remarkably homogenous species. Now, however, the introduction of hatchery-produced fish could seriously upset the balance of the wild stock by interbreeding, thereby altering the genetic composition of the wild stock and reducing the homing instincts in the offspring.

Conservation of fresh waters

Trout streams are probably subject to more careful management than other stretches of water but the principles adopted are worthy of use in managing most waterways.

Silting is one of the greatest problems and its effects are felt first by the disappearance of weed beds needed to produce a good population of invertebrates upon which the fish feed. In a healthy trout stream, areas of water crowfoot support a large invertebrate fauna, but the crowfoot can disappear rapidly if silt builds up to impede flow and make the water too

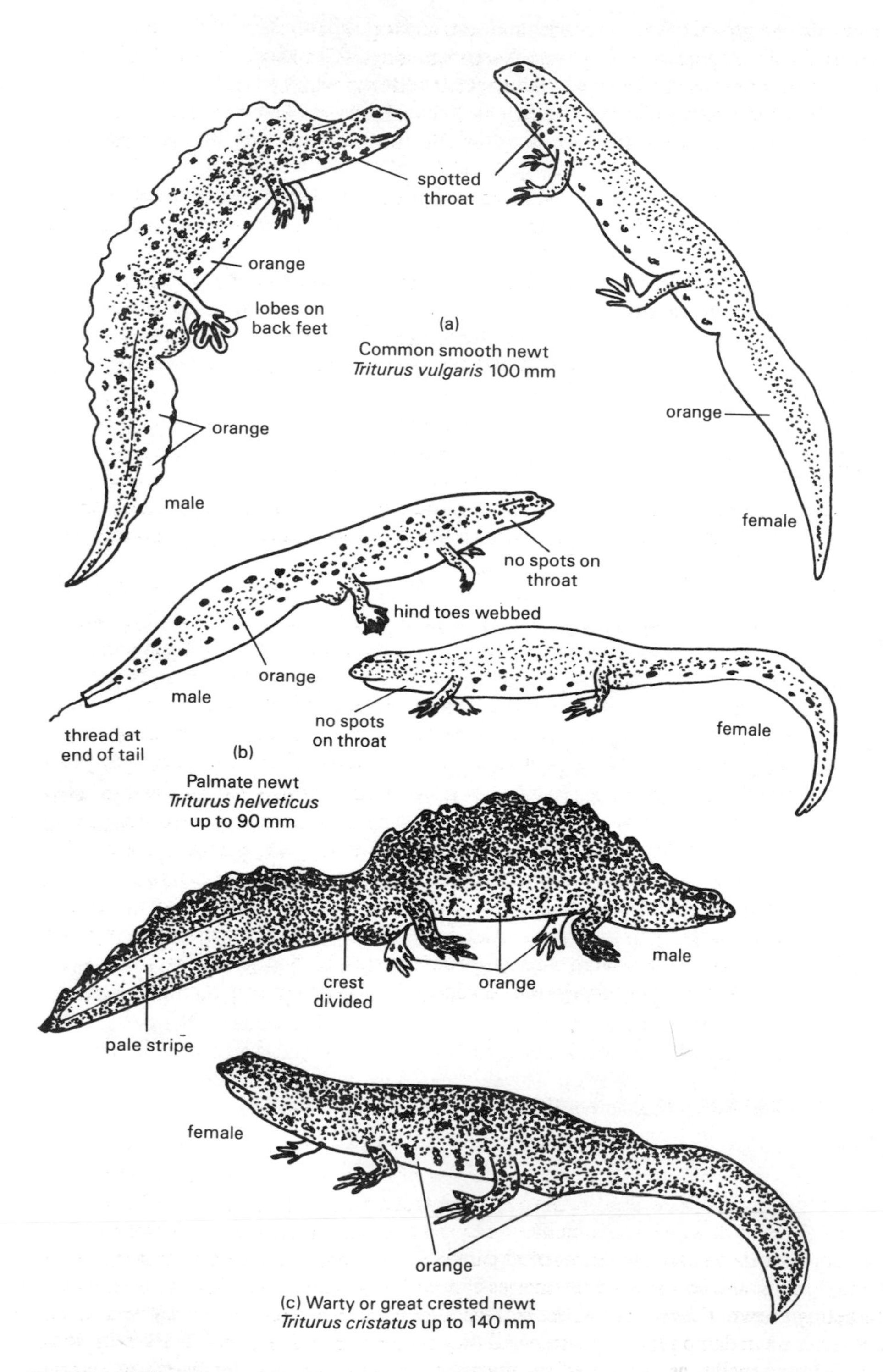

Fig. 13.4 The males and females of the three British species of newt to show distinguishing features

shallow for its growth. Water starwort and watercress quickly overgrow the crowfoot and increase the silting process. Yellow flag often grows along the banks of a river and will remain within its confines unless there is a build-up of silt within the flag bed and beyond, when the flags will increase to exploit these conditions. Such a build-up of silt will have its effect by increasing other areas of silt, resulting in the further extinction of areas of water crowfoot.

Indiscriminate dredging is often employed to cope with the situation, further aggravating the problem by the removal of large areas of river bed, thereby destroying its structure and flow-pattern.

A recent development in the clearance of silt, without the use of machinery but by natural means, could be one answer. It consists of introducing large quantities of microscopic fossilized marine plants, called coccoliths, together with the chalk in which they were embedded. Under normal conditions a silt bank, high in organic content, is almost a solid mass saturated with water. The current pushes the mass along to pile it up elsewhere. Bacterial decay of the organic matter cannot operate under the waterlogged conditions. By virtue of their small size, the coccoliths, in a treated stream, penetrate the silt and release the water, permitting oxygen to enter so that the normal processes of decay can take place, liberating food for detritus feeders such as worms and small crustaceans. This break-up and release of organic matter enhances the ecology of the steam instead of the reverse being wrought by mechanical dredging.

Shading by bankside trees can occlude light falling on weed beds. By controlling the amount of shading, excessive growth of water weed can be prevented, reducing the need for weed-cutting. Fish require a certain amount of cover, however, and weed provides both shelter and food for other organisms, so light should not be reduced by more than 50 per cent. Planting of bankside trees in unshaded areas should take into account other operations such as the necessary cutting of weeds in areas where there is overgrowth. Tree planting is often necessary to preserve the banks and of course their shading effect will depend upon the direction in which the stream or river lies. Narrow streams usually receive enough shade from bankside herbage to give a fair measure of weed control. However, banks can quickly be broken down and plants destroyed by cattle allowed unrestricted access to a stream. This can be overcome by the construction of fenced-in areas where cattle can drink.

Finally, it must be remembered that a meandering steam offers a great variety of habitats, such as swift-and slow-flowing reaches, riffles, and different areas of substrata ranging from pebbles to mud, as well as a large variety of bankside plants. Straightening of streams to increase flow and the uncontrolled use of dredgers can drastically alter their ecology.

Conservation of rare species

The conservation of rare species should rightly be an important part of the management of fresh water. Yet the term 'endangered species' is difficult to define, especially when it is applied to invertebrates which do not necessarily rank high in the hierarchy of those to be conserved, or of those which catch the eye of the relatively uninitiated. It is the old story of otters immediately the favourites of all animal lovers, which receive the most attention.

Frogs, toads, and two of our three species of newt (the smooth newt, *Triturus vulgaris* and the palmate newt. *T. helveticus*) leave the water after metamorphosis and spend the next two to three years in damp places on land until they become sexually mature. Some may make their way annually, as non-breeding juveniles, to water during the breeding season, returning to land along with the breeding adults for the rest of the year. Indeed, for the first time, special signs depicting a toad were erected in 1984 on roads at sites known to be toad crossings in the breeding season.

The warty or great crested newt, *T. cristatus*, spends all of its life in water, requiring deeper pools in which to live and breed. Until the 1950s it was to be found in a number of ponds throughout Great Britain but the filling in of ponds for agricultural or building purposes has greatly reduced its numbers to the point where it is now protected and it is an offence in law to destroy or remove specimens.

Figure 13.4 shows all three species of British newts in their breeding condition. The colours, especially of the males, become very bright and the males of *T. vulgaris* and *T. cristatus* grow larger, and quite flamboyant, crests. During the non-breeding season, *T. vulgaris* and *T. helviticus* become relatively insignificant in appearance and are often mistaken for lizards when on land. All of these amphibians are at risk from pesticides and other toxic substances, whether they are in the water or on land, since they feed on slugs, snails, worms, or other invertebrates which may carry these poisons.

Some conservationists may be concerned solely with the preservation of endangered species, but the conservation of fresh water, whether it be rivers, lakes, ponds or their catchment areas, is a matter of the greatest importance to all who use them. The maintenance of a correct balance of nutrients and the control of pollution in all its forms are vital to the continued existence of our waterways and for the survival of their inhabitants. Upsetting the balance, for possible short-term gain, can have long-lasting effects which may be difficult or even impossible to reverse.

Perhaps the best success story of recent years is that of cleansing the River Thames. It was once one of our most polluted rivers, receiving pollutants of every kind from industry, agriculture, and sewage. Active steps were initiated to reclaim, over a period of time, the river's purity. It has now been transformed into a body of water which is not only drinkable but which supports a healthy population of salmon and other river fish sensitive to pollution.

References

13.1 Carpenter, K.E. (1924) 'A study of the fauna of rivers polluted by lead mining in the Aberystwyth district of Cardiganshire', *Ann. appl. Biol.*, **9**, 1–23

13.2 Chandler, J.R. (1970) 'A biological approach to water quality management', *Wat. Poll. Control* **69**, (4), 415–22

13.3 Hynes, H.B.N. (1960) *The Biology of Polluted Waters*, Liverpool University

13.4 Phelps, E.B. (1944) *Stream Sanitation*, New York

13.5 Spence, K. (1984) 'Killer from the clouds: Time for action on acid rain', *Country Life*, 1493–98

13.6 Sutcliffe, D.W, (1983) 'Acid precipitation and its effects on aquatic systems in the English Lake District (Cumbria)', *An. Rep. Freshw. Biol. Assoc.*, 30–57

Bibliography

The titles included below do not aim to cover each section exhaustively but to be a useful list of those most widely in use. Many of these books contain details of papers, and other publications, which can be consulted by those wishing to study particular aspects in greater depth.

General ecology

Andrewartha, H. G. (2nd edn 1970) *Introduction to the Study of Animal Populations*, Chapman & Hall

Darlington, A. and Leadley Brown, A. (1975) *One Approach to Ecology*, Longman

Elton, C. (1958) *The Ecology of Invasions by Animals and Plants*, Methuen

Macfadyen, A. (2nd edn 1963) *Animal Ecology, Aims and Methods*, Pitman

Odum, E. P. (3rd edn 1971) *Fundamentals of Ecology*, Saunders

Sands, M. K. (1978) *Problems in Ecology*, Mills & Boon

Freshwater ecology

Clegg, J. (3rd edn 1965) *The Freshwater Life of the British Isles*, Warne

Edmondson, W. T. (ed), Ward, H. B. and Whipple, C. C. (2nd edn 1959) *Fresh-water Biology*, Wiley

Hynes, H. B. N. (1971) *The Ecology of Running Waters*, University of Liverpool Press

Macan, T. T. and Worthington, E. B. (1951) *Life in Lakes and Rivers*, Collins

Macan, T. T. (1963) *Freshwater Ecology*, Longman

Macan, T. T. (1973) *Ponds and Lakes*, Allen & Unwin

Miall, L. C. (1902) *The Natural History of Aquatic Insects*, Macmillan

Sterry, P. (1982) *Pond Watching*, Hamlyn

Townsend, C. R. (1980) *The Ecology of Streams and Rivers*, Studies in Biology No. 122, Arnold

Whitton, B. A. (1975) *River Ecology*, Blackwell

Ecological methods

Bishop, O. N. (3rd edn 1981) *Statistics for Biology*, Longman

Campbell, R. C. (2nd edn 1974) *Statistical Methods in Biology*, Cambridge

Elliott, J. N. (1971) Some Methods for the Statistical Analyses of Samples of Benthic Invertebrates, *Sc. Pub. Freshwat. Biol. Assoc.*, No. 25

Heath, O. V. S. (1970) *Investigation by Experiment*, Studies in Biology No. 23, Arnold

Lewis, T. and Taylor, L. R. (1967) *Introduction to Experimental Ecology*, Academic Press

Mills, D. H. (1972) *An Introduction to Freshwater Ecology*, Oliver & Boyd

Parker, R. E. (2nd edn 1980) *Introductory Statistics for Biology*, Studies in Biology No. 43, Arnold

Southwood, T. R. E. (2nd edn 1978) *Ecological Methods*, Chapman & Hall

Wratten, S. D. and Fry, G. L. A. (1980) *Field and Laboratory Exercises in Ecology*, Arnold

Identification

Freshwater plants and animals

Clegg, J. (3rd edn 1980) *The Observer's Book of Pond Life*, Warne
Frazer, D. (1983) *Reptiles and Amphibians in Britain*, Collins
Leadley Brown, A. M. (1970) *Key to Pond Organisms*, Nuffield Advanced Biological Sciences, Penguin
Macan, T. T. (1959) *A Guide to Freshwater Invertebrate Animals*, Longman
Mellanby, H. (6th edn 1963) *Animal Life in Fresh Water*, Methuen
Quigley, M. (1977) *Invertebrates of Streams and Rivers, A Key to Identification*, Arnold

Flatworms

Reynoldson, T. B. (2nd edn 1978) A Key to the British Species of Freshwater Triclads, *Sci. Pub. Freshwat. Biol. Ass.*, No. 23

Rotifers

Donner, J. (1966) *Rotifers*, Warne

Annelids

Brinkhurst, R. O. (1963) A Guide for the Identification of British Aquatic Oligochaeta, *Sci. Pub. Freshwat. Biol. Ass.*, No. 22
Elliott, J. M. and Mann, K. H. (1979) A Key to the British Freshwater Leeches, *Sci. Pub. Freshwat. Biol. Assoc.*, No. 40

Nematodes

Goodey, T. (1951) *Soil and Freshwater Nematodes*, Methuen

Crustaceans

Gledhill, T., Sutcliffe, D. W. and Williams, W. D. (1976) A Key to British Freshwater Cladocera, *Sci. Pub. Freshwat. Biol. Assoc.*, No. 5
Harding, J. P. and Smith, W. A. (2nd edn 1974) A Key to the British Freshwater Cyclopid and Calanoid Copepods, *Sci. Pub. Freshwat. Biol. Assoc.*, No. 18
Scourfield, D. J. and Harding, J. P. (3rd edn 1966) A Key to the British Species of Freshwater Cladocera, *Sci. Pub. Freshwat. Biol. Assoc.*, No. 5

Insects

Disney, R. H. L. (1975) A Key to the Larvae, Pupae and Adults of the British Dixidae, *Sci. Pub. Freshwat. Biol. Assoc.*, No. 31
Edington, J. M. and Hildrew, A. G. (1981) Caseless Caddis Larvae of the British Isles, *Sci. Pub. Freshwat. Biol. Assoc.*, No. 43
Elliott, J. M. (1971) A Key to the Larvae and Adults of British Freshwater Megaloptera and Neuroptera, *Sci. Pub. Freshwat. Biol. Assoc.*, No. 35
Hammond, C. O. (1977) *The Dragonflies of Great Britain and Ireland*, Curwen
Hickin, N. E. (1952) *Caddis*, Field Study Books, Methuen
Hynes, H. B. N. (3rd edn 1977) A Key to the Adults and Nymphs of the British Stoneflies (Plecoptera), *Sci. Pub. Freshwat. Biol. Assoc.*, No. 17
Linssen, E. F. (1959) *Beetles of the British Isles*, Warne
Macan, T. T. (2nd edn 1965) A revised Key to the British Water Bugs (Hemiptera-Heteroptera), *Sci. Pub. Freshwat. Biol. Assoc.*, No. 16

Macan, T. T. (1973) A Key to the Adults of the British Trichoptera, *Sci. Pub. Freshwat. Biol. Assoc.*, No. 28

Macan, T. T. (3rd edn 1979) A Key to the Nymphs of British Ephemeroptera, *Sci. Pub. Freshwat. Biol. Assoc.*, No. 20

Southwood, T. R. E. and Leston, D. (1959) *Land and Water Bugs of the British Isles*, Warne

Molluscs

Janus, N. (1965) *Land and Freshwater Molluscs*, Burke

Macan, T. T. (4th edn 1977) A Key to the British Fresh- and Brackish-water Gastropods, *Sci. Pub. Freshwat. Biol. Assoc.*, No. 13

Fish

Bagenal, T. B. (ed. 1970) *The Observer's Book of Freshwater Fishes*, Warne

Maitland, P. S. (1972) Key to British Freshwater Fishes, *Sci. Pub. Freshwat. Biol. Assoc.*, No. 27

Muus, B. J. and Dahlstrøm, P. (1971) *Collins Guide to the Freshwater Fishes of Britain and Europe*, Collins

Amphibians

Arnold, E. N., Burton, J. A. and Ovenden, D. W. (1978) *Reptiles and Amphibians of Britain and Europe*, Collins

Glossary

Allochthonous Pertaining to plant or animal products reaching an aquatic community from outside
Anadromous Fish which migrate from the sea to fresh water to spawn
Analysis of variance A procedure used for comparing more than two means
Autochthonous Pertaining to plant or animal materials produced within a body of fresh water
Autotrophic The ability to utilize the light energy of the Sun through photosynthesis to manufacture food, as in green plants and some bacteria (see **heterotrophic**)

Benthic region The area of bottom sediments in a lake or pond extending from the shore to deeper parts
Benthos Organisms associated with the bottom sediments
Biomass The weight (usually expressed as dry weight) of living material per unit area
Boundary layer Thin layer of water flowing over a stone in a stream bed

Catadromous Fish which migrate from fresh water to the sea to spawn
Community A group of populations which interact with each other (see **population**)

Density The number of individuals per unit area
Detritivores Animals which feed upon detritus
Detritus Organic matter which is in the process of decomposition
Discharge The volume of water flowing in a stream at a particular point

Ecological niche The position occupied by an organism within a community relative to others in the community
Ecosystem The situation arising from the interaction of a number of organisms with their environment and with each other
Elytra (sing. **elytron**) In insects, the hardened, chitinized wing cases
Epilimnion The warm upper layer of water in a lake, between the surface and the thermocline
Epineustic See **neuston**
Epiphytic Living wholly, but not parasitically, on other plants
Eutrophic Water containing abundant nutrients which produce a rise in productivity and cause a lack of oxygen in lower areas

Food chain The passage of food through populations within a community
Food web The relationships between a number of food chains

Gross production The total amount of light fixed by green plants, through photosynthesis, in a given period of time

Habitat The place where an animal or plant lives
Heterotrophic The use, by herbivores and carnivores, of complex organic materials as food (see **autotrophic**)
Homeostasis Maintenance of constant internal conditions in a fluctuating environment

Hydrofuge Water repellent
Hydrosere Plant succession from water to land
Hypolimnion The cold lower layer of water in a lake, between the thermocline and the bottom
Hyponeuston Organisms living in association with the underside of the surface film
Hyporheos Community of small animals living in the interstices of coarse sediments in a stream bed

Index of diversity A parameter used to quantify the diversity of a community
Invertebrate drift Movement of invertebrates, either up or down a stream or river

Leaching Removal of soluble compounds in the soil by water, usually rain
Lentic Associated with still water
Lithophilous Living associated with stones
Lotic Associated with moving water

Microhabitat A small part of a habitat with its own characteristic features

Nekton Organisms moving actively between the pelagic and benthic regions
Net primary productivity The amount of organic matter stored in plant tissues in excess of the respiratory needs of the plant
Net production The gross production minus the amount of energy required for respiration. That which remains is available for growth
Neuston Organisms living on the upper side of the surface film

Osmoregulation The control of osmotic pressure

Parasite An organism which lives on the living tissues of another, usually to the latter's detriment
Pelagic region Area of water which is free from the influence of the shore or the bottom regions of a lake or pond
Perennation Method of overwintering
Periphyton See **epiphytic**
Photosynthesis In plants, the combination of water and carbon dioxide, in the presence of chlorophyll and light, to form sugar, with the evolution of oxygen
Plankton Drifting, very small, aquatic plants (phytoplankton) and animals (zooplankton)
Poikilotherm An animal whose body temperature is dependent on the external temperature
Population A group of organisms of the same species occupying a particular ecological habitat (see **community**)
Primary producers Organisms which are capable of synthesizing their food from simple substances through photosynthesis or chemosynthesis
Productivity The amount of material produced per unit time
Standing crop The amount of living material present in a community at a particular moment
Stolon Creeping branch which is capable of giving rise to new individuals
Succession Progressive change in the plant and animal life of a community, from the first colonists to climax conditions

Thermocline Zone of a lake, between the epilimnion and the hypolimnion, where there is a rapid change in temperature

Transect Method of measuring the distribution of plants and recording the results graphically. A line transect gives a profile of vegetation and a belt transect shows horizontal distribution within parallel lines, usually 1 m apart

Trophic Pertaining to nutrition

Trophic level The level in a food web at which an organism, or a group of organisms, occurs

Turion An overwintering organ formed by a number of aquatic plants

Index of common names

References to figures, plates and tables are printed in bold type: e.g. **8.2a, plate 8.3, table 8.4.**

General index

References to figures, plates and tables are printed in bold type: e.g. **8.2a, plate 8.3, table 8.4.**